Ethics and the Environmental Health Profession

Providing an exploration of the discussion on how a stronger ethical basis for the work of environmental health practitioners (EHPs) can contribute to improved public health, because EHPs/environmental health officers (EHOs) come into daily contact with members of the public to address threats to their health and wellbeing. This book examines what is meant by professional ethics and the role of professional bodies in ensuring members of the profession act ethically. They should be expected to act as guardians of the public interest, without fear or favour. Ethics is an integral part of public health decision-making and needs to be incorporated into public and environmental health policy development and decision-making. This work, while seeking to stimulate debate within the environmental health profession, will examine what this means for EHPs.

Ethics and the Environmental Health Profession: The Importance of Being Ethical explores the process of ethical decision-making in the environmental health profession and asks the question of whom EHPs have an ethical duty towards in their work. It looks at a variety of ethical issues which arise for EHPs working in local government, as consultants or for commercial companies.

This book is useful for EHPs globally, other local government officers, educational establishments where environmental and public health courses are run globally, elected members of local authorities, and policymakers at national and local government levels.

Stephen Battersby, MBE, is a freelance environmental health and housing consultant, and researcher. He is a fellow of the Chartered Institute of Environmental Health (CIEH) and the Royal Society for Public Health (and is now retired). He is a past President of the CIEH and is currently a Vice President.

David Miles, Hon DUniv (Kingston), FRSA, joined Kingston University as a lecturer in 1972. He was Professor of Business Education in 1993, Dean of the Faculty of Business and Law from 1983 to 1999, and Pro Vice-Chancellor External Affairs from 1999 to 2007. He maintains contact with the University as Honorary Dean of Alumni and Development and is currently Chair of Shenehom Housing Association Ltd – a care home for people with enduring severe mental health issues.

Routledge Focus on Environmental Health

Series Editor: Stephen Battersby, MBE PhD, FCIEH, FRSPH

Tackling Environmental Health Inequalities in a South African City?
Rediscovering Regulation, Local Government and its Environmental Health Practitioners
Rob Couch

Leadership Lessons from a Global Health Crisis
From the Pandemic to the Climate Emergency
Jo Nurse

Sewerage and Sewage as an Environmental Health Issue
Stephen Battersby

Private Sector Housing and Health
Evaluating the Effectiveness of Regulation Intended to Protect the Health of Tenants
Paul Oatt

Dampness in Dwellings, 2nd Edition
Causes, Effects and Remedies
Véronique Ezratty and Stephen Battersby

Local Air Quality Control
Theory and Practice
Mark Hardwick, Nahum Kidan, Steve Manning and Nathaniel Smith

Ethics and the Environmental Health Profession
The Importance of Being Ethical
Stephen Battersby and David Miles

Ethics and the Environmental Health Profession

The Importance of Being Ethical

Stephen Battersby and David Miles

Routledge
Taylor & Francis Group
LONDON AND NEW YORK

First published 2026
by Routledge
4 Park Square, Milton Park, Abingdon, Oxon OX14 4RN

and by Routledge
605 Third Avenue, New York, NY 10158

Routledge is an imprint of the Taylor & Francis Group, an Informa business

British Library Cataloguing-in-Publication Data
A catalogue record for this book is available from the British Library

ISBN: 978-1-041-03117-8 (hbk)
ISBN: 978-1-041-03780-4 (pbk)
ISBN: 978-1-003-62533-9 (ebk)

DOI: 10.1201/9781003625339

Typeset in Times New Roman
by codeMantra

“Having knowledge of an unethical act and allowing it to continue can spread a contagion that can affect multiple beings in society”

Bertrand Russell (1872–1970)

Contents

Series preface

This series is ever expanding, illustrating the issues relevant to environmental health. The aim remains as ever, to explore environmental health topics traditional or new, and sometimes raise contentious issues in more detail than might be found in the usual environmental health texts. It is also a means whereby environmental health issues can be discussed with a wider audience in mind.

This series is now an important part of the professional landscape, as is apparent from the titles published so far. Environmental health practitioners bring their expertise to a range of situations and are deployed differently but not always to the best effect so far as public health is concerned. All too often politicians both at the national and local levels are unaware of what is environmental health, and what practitioners do, how they work, and indeed if they are aware how they can be assured of their competence. Invisibility is an issue for the profession, and this series is an attempt to raise the profile and highlight the work of environmental health practitioners. This series provides a means for practitioners to improve the profile of the profession as well as provide a source of information for others.

We want to encourage readers and practitioners, particularly those who might not have had work published previously. I am particularly keen that this series is seen as an opportunity for first-time authors and as ever urge students at whatever level (to consider this an avenue for publishing findings from their research as some have already done). The hard work that has gone into a dissertation or thesis should not be unused or forgotten. We can provide advice on turning a thesis into a book. This series can be a way of extending a presentation, paper, or training materials, so that they can reach a wider audience. We know that practitioners do not have time to write a monograph but the series can accommodate multi-author editions and so a team could write up their work, thereby lightening the burden on an individual.

This series has the advantage of having a relatively quick turnaround from submission of the manuscript to publication and can be more up to date and immediate than a standard textbook or reference work. It is also a route for analysing actions, and reporting on what worked in practice, what was successful, what wasn't, and why so can provide useful insights for others working in the field.

Contributing to this series fulfils the essence of CPD as it requires reflection, analysis, and acquiring new skills and knowledge; what any good professional should do. I am pleased to be working with Routledge to provide this opportunity for practitioners.

It is not intended that this series always takes a wholly "technical" approach but provides an opportunity to consider areas of practice in a different way, for example, looking at the social and political aspects of environmental health in addition to a more discursive approach on specialist areas.

Our hope remains that this is a dynamic series, providing a forum for new ideas and debate on environmental health topics. If readers have any ideas for titles in the series, please do not be afraid to submit them to me as series editor via the e-mail addresses below.

Forthcoming monographs are likely to cover such topics and water quality and global environmental health, and we are looking at including in the series a monograph on a conference on housing and health in extreme events.

We welcome suggestions from any practitioners in the environmental health field (not just environmental health officers [practitioners]) wherever in the world for any new topics or publications in the series. I hope readers with ideas for a monograph will get in touch via Ed.Needle@tandf.co.uk or Martha.Luke@tandf.co.uk.

Stephen Battersby MBE PhD, FCIEH, FRSPH
Series Editor

Introduction

Environmental health as a profession began with ethical purposes, and its founders unashamedly sought to make a better world for all citizens. These days the profession is under many pressures, and there is no doubt that it might be thought by some that "ethics" is something of a luxury and pragmatism overrides such considerations. However, ethics are something that are essential to any group of people who consider themselves a profession.

We consider in this book what makes a profession and focuses on professional ethics as applicable to environmental health professionals. History (as we will see in Chapter 1) bears out the impact of this ethical imperative, lives have been saved, and conditions of life have improved. Practitioners, wherever they work, have an obligation to articulate and update the underlying ethical principles of the profession to the benefit of wider society, the good standing of the profession and themselves as individuals, and improvement of the services they provide. This is not just a matter for their professional associations.

The idea for this book came about as the result of debates at annual meetings of the Chartered Institute of Environmental health (CIEH). While motions under discussion did not mention ethics, it seemed that there were some ethical questions. Some members felt their professional body needed to be better focussed on restoring the CIEH's professional identity. They thought this had become diluted in recent years as the executive strove to expand the membership base. This meant to some that the CIEH's role as a professional body, representative of those holding a recognised environmental health qualification, had become confused. It was also thought that alterations made to the practical and professional training requirements and their assessment had led to a proportion of employers believing the qualification no longer provided the holistic skills in the core areas of environmental health practice needed for a flexible workforce that is also part of the public health workforce. However, while much of the work refers to professional bodies in the UK much of the discussion is relevant to EHOs/EHPs in other countries. There is also reference to other associations, and the information has been taken from their websites and is publicly available.

Within the UK and indeed around the world what in the past we have considered as environmental health officers or EHOs (or indeed Public Health Inspectors – PHIs), now may be called environmental health practitioners or EHPs or some other variant. This is recognised, so we have used EHO/EHP throughout this book to cover all titles.

Whatever their title or designation their work brings them into close contact with members of the public in various settings. They then must make decisions that affect individuals and their communities. Such decisions must have ethical implications and cannot be left to an individual's notion of morality.

The aim of this book (which is a polemic rather than a deeply researched academic text) is to stimulate further debate within the environmental health profession including the professional associations. We argue that considering a stronger ethical basis for the work of EHPs (or EHOs) can contribute to improved public health. There is a question, to whom does the EHP have an ethical duty at any one time – the law, employers (public and/or private), clients, tenants, consumers, the practitioners themselves, the "truth"? How do you resolve conflicts between them and what support is there when making such decisions?

David Miles and Stephen Battersby

About the authors

Dr Stephen Battersby, MBE FCIEH FRSPH, is a freelance environmental health and housing consultant, and researcher. He is a fellow of the Chartered Institute of Environmental Health (CIEH) and the Royal Society for Public Health (and is now retired). He is a past President of the CIEH and is currently a Vice President.

He is also a visiting senior research fellow at the Centre for Environment and Sustainability at the University of Surrey, where over the years he has undertaken research and teaching roles. He has also been an associate of the Safe and Healthy Housing Unit at Warwick University.

Over the years he has worked with Parliamentarians both on research and on campaigns to improve legislation on environmental health topics.

He was the editor of *Clay's Handbook of Environmental Health* for three editions (20th–22nd) and has also had a number of academic papers published as well as contributing to a number of titles in the Routledge Focus on Environmental Health Series. Orcid: https://orcid.org/0000-0003-4564-6561

Professor David Miles Hon DUniv (Kingston), FRSA, joined Kingston University as a lecturer in 1972. He was Professor of Business Education in 1993, Dean of the Faculty of Business and Law from 1983 to 1999, and Pro Vice-Chancellor External Affairs from 1999 to 2007. He maintains contact with the University as Honorary Dean of Alumni & Development.

David was a City Councillor in Westminster (1974–1982), a member of the Education Committee of the Inner London Education Authority (1976–1981), and a Labour Candidate in the 1979 General Election.

His other roles have included: Chair of the Council for Education in World Citizenship; Trustee of the Citizenship Foundation; Deputy Lieutenant of Greater London; Trustee of the Institute of Direct Marketing; and Trustee of Kingston University Students' Union. He is Chair of Shenehom Housing Association Ltd – a care home for people with enduring severe mental health issues.

Acknowledgements

The authors wish to acknowledge the great help provided by colleagues in preparing this book and offer our thanks. There have been discussions with representatives of both the CIEH and REHIS, which were particularly helpful.

We also wish to express our gratitude to those colleagues who submitted their thoughts on ethical dilemmas and issues that we include in Chapter 4 and offered feedback on the drafts that we then prepared for inclusion. Their help is greatly appreciated. The submissions and testimonies in Chapter 4 have given real meaning and grounding to our thoughts in the other chapters. We invited these submissions on the basis that they would be anonymised and so we cannot name them here.

We should also acknowledge and thank Cynthia Battersby who read through the final drafts and offered invaluable support and assistance.

Stephen Battersby and David Miles

Acronyms and abbreviations used

APHI	Association of Public Health Inspectors (which became CIEH)
AQMA	Air Quality Management Area
BICoF	Building Inspector Competence Framework
BoC	Board of Certification (Canada)
BSR	Building Safety Regulator
CEUs	Continuous Education Units (CEUs) (South Africa)
CEnvH	Chartered Environmental Health practitioner
CIEH	Chartered Institute of Environmental Health
CIOB	Chartered Institute of Building
CIPD	Chartered Institute of Personnel and Development
CIPHI	Canadian Institute of Public Health Inspectors
COPD	Chronic obstructive pulmonary disease
CPD	Continuing Professional Development
CPSM	Council for the Professions Supplementary to Medicine (Jamaica)
CTSI	Chartered Trading Standards Institute
DPH	Director of Public Health
EHA	Environmental Health Australia
ECHR	European Convention on Human Rights
EHO	Environmental Health Officer
EHP	Environmental Health Practitioner
EHRB	Environmental Health Registration Board
EWS1	External Wall Cladding Assessment
FPH	Faculty of Public Health
GMC	General Medical Council
GP	General Practitioner
HHSRS	Housing Health and Safety Rating System
HMO	House in Multiple Occupation
HPCSA	Health Professions Council of South Africa
HPT	Health Protection Team
IDEA	Inter-Disciplinary Ethics Applied (IDEA) Centre (Leeds University)
IFE	Institute of Fire Engineers
JAPHI	Jamaica Association of Public Health Inspectors

LGO	Local Government Ombudsman
LHA	Local Housing Authority
MOH	Medical Officer of Health
NCDs	Noncommunicable diseases
NEHA	National Environmental Health Association in the USA
OED	Oxford English Dictionary
PARN	Professional Associations Research Network
PHI	Public Health Inspector
RBH	Rochdale Boroughwide Housing Association
RBI	Registered Building Inspector
REHIS	Royal Environmental Health Institute of Scotland
REnvH	Registered Environmental Health Practitioner
SAIEH	South African Institute of Environmental Health
TMO	Tenant Management Organisation
UKHSA	United Kingdom Health Security Agency
UKPHR	UK Public Health Register
WASC	Water and Sewerage Company
WWA	Welfare Workers Association

Table of cases

1 Aims and background

Ask you what provocation I have had?
The strong Antipathy of Good to Bad.
When Truth or Virtue an Affront endures,
Th' Affront is mine my Friend, and should be yours.

Epilogue to the Satires, Dialogue II, lines 197–200,[1] Alexander Pope poet and satirist 1688–1744

This is Pope's response to those who criticised his preoccupation with ethical behaviour.

Introduction and aim of this book

Environmental health as a profession began with ethical purposes. Those on whose foundations the profession is built had roots in honest scepticism, knowledge-based actions coupled with a concern for humanity believing always that health can be improved and ill health prevented. There have also always been vested interests that try to prevent such improvements.

That is why this book aims to set out what is meant by ethics in general and professional ethics in particular and why this remains a key but often a forgotten issue for environmental health officers (or practitioners) (EHOs/EHPs). Why is this important? EHPs deal with the public every day, and it is fundamental that there should be an ethical basis for any advice given and any responses to questions need to be trusted as honest in purpose; founded on disinterested facts and professional knowledge. EHOs/EHPs are part of the public health workforce, and ethics is central to public health policy and practice but as has been said "it should not be presumed that all moral values will be equally shared by every public health policy maker or practitioner", and an understanding of ethics should be considered a key competency for people working in public health.[2] The Faculty of Public Health has reinforced this, saying "Public health is grounded in values and moral norms that guide decisions, behaviours, policy and practice. Ethics and law are accordingly core public health competencies for professional practice".[3]

DOI: 10.1201/9781003625339-1

This book will consider why an effective code of ethics for the profession as a whole is a pre-requisite for maintaining and enhancing the reputation of the profession when much of the world has changed and indeed continues to change with increasing rapidity. EHPs wherever they practise should take pride and gain fulfilment in their careers from identifying the ethical foundation and that being a member of the profession is an honourable role in life. We therefore examine different codes of ethics and how they are applied and what can be learned.

There is a concern that despite the threats to health from environmental stressors including the climate emergency, the environmental health profession seems to be undervalued (and indeed invisible) particularly in England. It has been said that environmental health is "doubly invisible, in that public health is culturally invisible, and within that sphere environmental health is invisible".[4] While some changes over the years might be down to politics and "deregulatory fervour" and economic pressures on local government, the professional body, the Chartered Institute of Environmental Health (CIEH) can also seem to be diminished in influence and services to members, so that it seems some members of the profession choose not to maintain their membership or others not join at all. It then becomes more difficult to maintain a high ethical standard within the profession.

The status of the term "profession" is not tightly defined but varies with the context. Some professions are highly exclusive. In the United Kingdom, for example, the right to plead on behalf of clients in the courts has been largely restricted to members of the Bar and there are controls on who can practise as a solicitor.[5] The right to practise medicine is restricted to persons registered with the General Medical Council. Other "professions" have much less privileged status including EHPs, yet since the 19th century arguably the public health movement including EHOs has probably prevented more illness, saved, and improved more lives than the medical profession. Thus, environmental health has been a major part of civic society, but it lacks the prestige and public recognition that it needs to meet the purposes of contemporary society.

It is forbidden for anyone who is not a member of the CIEH to use the title "Chartered Environmental Health Practitioner" or "Registered EHP",[6] but there seem to be few, if any circumstances where membership of the Institute is a prerequisite, though employers could specify membership of the Institute as a condition of appointment. That said, membership in the correct grade is now required to gain and maintain CIEH registration. However, it is possible and with no legal impediment, for a person to complete an environmental health degree and then call themselves an EHP without being registered, so long as they did not use either of the protected titles. Although unsurprisingly, the CIEH would discourage this.[7]

As is pointed out later in this book, in Scotland there is a requirement for every council to employ a suitably qualified EHO. That leads to the next issue that in the context of the environmental health profession apart from in Scotland, which is the status of EHOs/EHPs since the reforms of 1974 prior to which the Chief EHO was a statutory appointment. There have been Chief EHOs at the government level and that remains true for Wales and Northern Ireland. The question might be, have these changes undermined the profession and particularly so in a period of economic uncertainty and what does this mean for professional ethics?

This book explains why ethics is fundamental to any profession and should be one factor that distinguishes EHOs or EHPs from people who may have little relevant qualification but are recruited to carry out tasks traditionally undertaken by a "fully qualified EHO". Furthermore, the CIEH Workforce Survey[8] (2021) found that because of shortages or delays in recruitment 87% of local authorities in England were relying on agency staff for environmental health work and 56% have had vacancies unfilled for more than six months. The data suggests that there are more professionals working to deliver environmental health services who are not fully qualified and/or trained in environmental health than there are fully qualified EHPs. Paradoxically in England and Wales, it seems the statutory status of the EHP has reduced while many powers and areas of responsibility have increased. The breadth of knowledge and transferrable skills of a fully qualified EHO/EHP should make the profession one that is recognised as crucial for dealing with the sometimes complex, problems and issues facing the public.

It is all too common nowadays that environmental health practitioners are managed by those with little knowledge or insight into how the work can safeguard or improve public health. Perhaps for managerial convenience, the work of the EHP/EHO has often become a tick box exercise with little scope for professional judgement based on training and education, knowledge and expertise, while at the same time being out on the street means they are making decisions beyond the control of management. Would the profession be less invisible (and perhaps more likely to encourage entry to the profession) with a stronger approach to ethics? The question then arises "quis custodiet ipsos custodes" – (roughly translated as "who will guard the guards themselves") and ensure ethical standards are maintained.

While environmental health is (or should be) concerned with health inequality there should also be an awareness of how the work of EHOs/EHPs fits in with the economy – particularly when it is argued that regulation inhibits economic growth (although it can also be a driver of innovation and contributor to a healthier workforce). In that context it is worth considering that Kate Raworth has argued that as economics "now guides the management of nations and our planetary household, profoundly influencing the lives of us all" asking "is it time for economists to get serious about ethics?".[9] She quotes George DeMartino approvingly "that when a profession seeks influence over others it necessarily takes on ethical obligations". That is true of EHOs/EHPs whether they recognise this or not.

We want to stimulate debate within the profession and to some extent this has been prompted by debates at recent CIEH general meetings. If "ethical questions" arise at all, how are they addressed by the professional bodies for EHOs and EHPs and by other professionals and their professional or regulatory agencies? We seek to examine the role and effectiveness of other professional bodies (not all related to environmental health) in ensuring ethical practice. When it comes to environmental health, are the professional bodies delivering the needs of society and communities such that the role and function of EHPs (or EHOs)[10] is adequately understood? This necessarily includes a brief look at registration and regulation of environmental professionals.

This book is largely written from the perspective of an environmental health professional with more than 50 years' experience in various arenas (but is not a moral philosopher) while looking to the future.

Professions and ethics

A theme of this book is the question of what makes a profession and a "professional" and whether a notion of ethics is essential at both the collective and individual level. Defining a profession has been the subject of number of papers over the years and how a profession is conceived and defined.[11]

Professionals can be said to be made via formal education, structured training that forms continuing professional development (CPD), and more informal unstructured in-service training. The Australian Council of Professions has defined a profession as a

> *disciplined group of individuals who adhere to ethical standards and who hold themselves out as, and are accepted by the public as possessing special knowledge and skills in a widely recognised body of learning derived from research, education and training at a high level, and who are prepared to apply this knowledge and exercise these skills in the interest of others.*[12]

It is therefore inherent in the definition that a code of ethics should govern the activities of each member of the profession requiring behaviour and practice beyond the personal moral obligations of any individual whether a member of the profession or not (what might be called general or global ethics). Such codes define and demand high standards of behaviour in respect to the services provided to the public and in dealing with professional colleagues. They are or should be enforced by the body that regulates the profession and that may be different in different countries. However this is achieved, the codes and the system for enforcement should be publicly available and accepted by wider society.

According to an article in *The Lancet,* one of the defining characteristics of professions "has always been that they have their own codes of ethics and are responsible for professional discipline. While professional ethics used to be thought of as something of a platitude, today that is changing".[13] We examine this in more detail in Chapter 2 looking particularly at the environmental health profession.

A brief history of social change and the development of environmental health – was there an ethical dimension?

Human activity has impacted the environment from the day fire was discovered and unwanted material was cast aside – waste. Prior to the 18th and 19th centuries smoke was seen as the major problem with complaints spanning many centuries. In the reign of Edward I, there was denunciation of "smoke nuisance" from the burning of coal.[14] Until the Industrial Revolution the problem was restricted to the major cities and towns. The urbanisation of the growing population brought many other public health problems and "nuisances" as well as increased smoke emissions from factory chimneys.

Why is there a need to consider history? Environmental health as we know it today grew out of the 19th-century sanitary and other reforming movements reflecting problems with poverty, unsafe water supplies, insanitary housing, and adulteration of food. It can be argued that the sanitary movement of the 19th century including improved water supplies and sewage treatment saved many more lives than medicine, which had relatively little impact on the lives of ordinary people, perhaps not until the advent of mass vaccination and antibiotics. While the work and possibly ethical considerations and pressures on EHOs and EHPs are very different in the 21st century, it is useful to look at what motivated those who brought about changes 150–200 years ago and what motivates EHOs (EHPs) today.

If EHPs are now "professionals", it is also worthwhile to see how the profession and the professional body have developed. If it is a requirement of professions that they have a code of ethics that recognises its obligations, then such a code has not always existed. The acceptance of the need for a code reflects the development of the profession, the educational level of members, and the role of the professional body(ies). There remains the question about the effectiveness of such codes of ethics.

Reformers in the 19th century

The history of the environmental health profession is covered in detail in other works.[15, 16, 17, 18] So far as we can tell the motivation to address the public health problems stemmed from several developments though the extent to which ethical considerations played a part varied and are not always clear. As a matter of convenience and pragmatism, the focus of this book is primarily on England.

The Age of Reason or The Enlightenment had been a philosophical movement in Europe from the 17th century through to the end of the 18th century. At its core was a belief in the use and celebration of reason or rationality, the power by which humans understand the universe and improve their own condition. The goals of rational humanity were considered to be knowledge, freedom, and happiness. It has been argued that The Enlightenment[19] ended with the French Revolution, although the thinking and principles continued to influence society. The Enlightenment produced secularised theories of psychology and ethics by thinkers such as John Locke and Thomas Hobbes giving rise to what were then considered radical political theories. It is worth considering at this stage that some of the reforms came about because of the fear of revolution and mob rule as well as outbreaks of diseases such as cholera.

Ethics is separate from religion or religious beliefs, but when we look at social history and developments leading to environmental health as we know it today, these were in part a reflection of religious beliefs. In the 19th century, illness came to be seen as an indicator of poor social and environmental conditions, as well as poor moral and spiritual conditions. Cleanliness was embraced as a path both to physical and moral health. Cleanliness, piety, and isolation were seen to be compatible and mutually reinforcing measures to help the public resist disease. At the same time, mental institutions became oriented towards "moral treatment" and cure.[20] As the result of the work of social reformers and campaigners plus increased knowledge about sanitation, society changed its thinking about public responsibility for citizens' health. Protecting health became a social responsibility with inspectors of nuisances then sanitary inspectors responsible for surveillance. Unfortunately, these people were working for medical practitioners who were better placed to claim the glory.

In the 18th and 19th centuries, we had the development of shareholder capitalism, whereby the wealthy could invest in ventures to make more money which with the expanding empire guaranteed a market for goods. Slaves were seen as just another commodity. Although there was government support for innovation, the Crown did not exercise the same level of control (or dead hand) as in France, for example, so that innovative technologies could be brought to market and brought into service much quicker.[21] There was also an ample and accessible supply of coal as a source of energy as well as other raw materials.

Along with developments in science and scientific and technological knowledge, the Victorian age was noted for the influence of religion on society, more than today. The Christian ethos had a major influence on several social reformers. It can be said that some of the issues that environmental health practitioners are dealing with today were first addressed in the 19th century, and so in some part but certainly not exclusively, it can be said that historically this might influence the ethics of environmental health as some of the examples here might illustrate.

The term Christian Socialism was first appropriated by a group of British men including Frederick Denison Maurice a theologian, and novelist Charles Kingsley who was also an Anglican clergyman, and others, who founded a movement that took shape in England immediately after the collapse of the Chartist movement in 1848. Their general purpose was to argue for "the Kingdom of Christ" its "true authority over the realms of industry and trade", and "for socialism its true character as the great Christian revolution of the 19th century". Marx had characterised religion as "the opiate for the people", Kingsley argued that the Bible had been wrongly used as "an opium-dose for keeping beasts of burden patient while they were being overloaded" and as a "mere book to keep the poor in order".

Octavia Hill was also a Christian Socialist, who pioneered improvements in housing, taking over management of properties, including repairs and maintenance, as well as rent-collecting. Anglican Josephine Butler campaigned successfully for the repeal of the Contagious Diseases Acts, and against child prostitution.[22] These provisions allowed police officers to arrest women suspected of being prostitutes in certain ports and army towns and other places. There was no set definition of prostitution, so all was left to the police officer's discretion. If a woman was declared to be infected, she would be confined in what were known as lock hospitals.

Anthony Ashley-Cooper 7th Earl of Shaftesbury was a leading figure in the evangelical movement of the Church of England. and was called the "Poor Man's Earl". He campaigned for better working conditions, reform to lunacy laws, education, and the limitation of child labour. He was associated with factory reform and shortening the working day in textile mills to ten hours. While serving as a member of the General Board of Health (1848–1854) and afterward, Shaftesbury insisted that the government sponsor new low-cost housing projects for urban workers and carefully inspect housing that already existed.

While religion and Christian values played a part in the work of many social reformers, this was a time of scientific development and religious beliefs were not the only driver of social reform, although perhaps there was not a strict demarcation. There were the utilitarians or Benthamites of which the most famous Victorian for environmental health practitioners was Edwin Chadwick. In the historical background, it is worth a few paragraphs on Chadwick as he was the first President of what is now the CIEH.

Utilitarianism had the idea that actions should be evaluated or assessed according to their consequences. It was opposed to the idea that people should pursue their own self-interest, and that some actions are inherently right or wrong regardless of their consequences.[23] The first systematic account of utilitarianism was developed by Jeremy Bentham, a philosopher and economist, although there had been earlier thinkers who were "theological". These included people such as Richard Cumberland (1631–1718) and John Gay (1699–1745). They believed that promoting human happiness was incumbent on us as a society because it was approved by God. Utilitarianism (discussed further in Chapter 2) is in essence a moral theory that judges actions based on their impact on human well-being. Bentham's version of utilitarianism, known as hedonistic utilitarianism, emphasised pleasure (for those who at the time had the vote).[11] To Benthamites the prescription for pauperism was deterrence. The pain of the workhouse should be so much greater than the pain of poverty and poor relief. It should not be forgotten that Chadwick's views on workhouses[24] were somewhat controversial because he believed in imposing a workhouse system that arguably was inhumane, as it separated families and lacked concern for individual liberties. As a follower of Bentham and the utilitarians, he was a believer in the welfare maximisation theory that argued for the greatest good for the greatest number (of those who were enfranchised).

In line with many other hagiographies of Chadwick, Vaishali has reinforced the EHPs' view that Edwin Chadwick[25] was a central figure in the 19th-century public health reform movement in Britain. This is true, but he was not without flaws. For a start there was his continued belief in the miasma theory (the belief that diseases were caused by a noxious form of "bad air" or miasma, which was thought to emanate from rotting organic matter), and his motivation was hardly altruistic. Hamlin[26] has provided a more balanced assessment. He demonstrates that Chadwick ignored medical reports showing that disease among Britain's working classes arose from a variety of conditions – destitution; bad housing and water plus sewage pollution along with long working hours in dangerous, toxic workplaces; and little education or training. To Chadwick the primary cause of pauperism and misery was not poverty or rampant capitalism, but filth. Clean water supplies and proper sewerage would not only alleviate disease but reform the manners of the lower classes and remove the danger of Chartism and the threat of revolution.

Nevertheless, Chadwick's work was certainly instrumental in the sanitary awakening, and revolutionised public health through the systematic improvement of urban sanitation and hygiene (doing the right thing for the wrong reason?). Best known for his, "The Report on the Sanitary Condition of the Labouring Population" of 1842 which highlighted the appalling conditions faced by the working classes. Chadwick's advocacy led to the passage of the Public Health Act of 1848, which established local health boards. His argument was also that a healthier working class would facilitate greater productivity in factories. The so-called "sanitary awakening" did however highlight the critical connection between environment and health.

Not all commentators have given such a positive endorsement as many, including Stewart,[27] who concedes that he was not an easy person to deal with, even as "the father of the modern environmental health profession" recognising that "health even as an issue for everyone". For instance, Ringen[28] attempted to place the origin of sanitary legislation in England, and Chadwick, within the overall dynamics of 19th-century social development. The public health movement was examined in the light of the transition of English society into the domination of the market ideology and the effect this had on health. He argued that the enactment of the 1848 Public Health Act was the unplanned reaction to the detrimental effects that the market ideology had on health in the industrial areas. He questioned the prevailing belief that sanitary reform was a humane contribution of public-spirited men and concluded that this state intervention was largely a necessity forced by the contradictions inherent in the market system. Ringen further suggests the utilitarians placed state interventions in the realm of the marketplace, so they "had to be effective by producing benefits and they had to result in future savings". While claiming to be scientific administrators (the first social engineers) and claiming to be objective, in reality they were the implicit defenders of the market ideology. Indeed, to support this he cites such diverse historians as Halevy and Hobsbawm on their agreement that the utilitarians were the most radical champions of the market ideology, for it provided them with the means to eliminate any social institution, tradition, or custom which might not satisfy their dictum. Responses to public health issues might also have reflected fears of revolution that was occurring across other European states – the Chartist movement in Britain[9] and the French Revolution and the Swing Rebellion were fresh in the minds of the British establishment and in addition the Paris Commune seized power in Paris in March 1871 and controlled parts of the city until May 1871.

In Mooney's review of Hamlin,[29] he highlights that Hamlin consistently singles out for praise someone that Chadwick attacked, the Edinburgh medical man William Allison. Alison subordinated economics to health rather than vice versa and empathised with those who were marginalised.

Looking at the work of Chadwick from almost 200 years later, some interesting questions of ethics arise. Hamlin has also argued elsewhere[30] when considering that medical practitioners were prescribing food to those in poverty and hungry (recognising the problems of hunger in poorer people) but to Chadwick and the Poor Law Commissioners, prescription of the basic commodities of life was unacceptable, "a violation of the laws of political economy". Hunger was seen as an incentive to work. The move to improve sanitation in 19th-century England and indeed Chadwick's 1842 report could be said to have further marginalised the urban poor and justified otherwise unacceptable governmental measures and social engineering. These were insidiously dangerous features of the sanitary movement according to Hamlin,[16] even though welcomed by liberal historians of the mid-20th century.

In addition to technology, science, religion, and philosophy as motors of change and of social reform, medical and epidemiological developments were also key. Important people and developments have been covered in detail, in other publications (and referenced above), but three are worth mentioning here to illustrate developments in the 19th century that illustrate the science and medical contribution to environmental health. These three are John Snow, William Duncan, and Thomas Fresh.

John Snow had seen cholera outbreaks first hand. He eventually undermined the miasma theory and the thinking of the General Board of Health, when in 1854 he was able to demonstrate a link between cholera and contaminated drinking water. He had argued that cholera was not airborne, but the germ theory was not developed at this time. Through his pioneering study, Snow (with the help of the reverend Henry Whitehead so even here religion played a part) identified the source of the outbreak as the contaminated public water pump on Broad Street (now Broadwick Street). He did this by mapping the deaths from cholera and noted that they were mostly people whose nearest access to water was the Broad Street pump. His studies of the pattern of the disease persuaded the local authority to disable the well pump by removing its handle. It was found that the water for the pump was polluted by sewage contaminated with cholera from a nearby cesspit. Most houses in the area had cesspits and were not yet connected to the sewerage system. Later, using statistics he was able to show that homes supplied by the Southwark and Vauxhall Waterworks Company, which took water from sewage-polluted sections of the Thames, had a cholera rate 14 times that of those supplied by Lambeth Waterworks Company, which took unpolluted water from the Thames upriver between Kingston on Thames and Esher at Seething Wells.

Snow made a further contribution to epidemiology in an 1857 pamphlet on the adulteration of bread as a cause of rickets. John Snow was a vegetarian and a teetotaller who campaigned for temperance societies but that does not explain his interests. He was clearly motivated to experiment and find solutions to problems affecting society.

The first Medical Officer of Health (MOH) in the country appointed by Liverpool Town Council in 1847 was Dr William Duncan.[31] Although Duncan did not produce great reports, he recognised that where people lived contributed to their ill health and that the town of Liverpool had health inequalities that needed to be addressed – the ill health of the Irish immigrants was due to where they lived; they were not unhealthy because of where they came from or who they were. Despite not publishing any great scientific treatise he secured improvements to public health in Liverpool working with the first Inspector of Nuisances (Thomas Fresh) who had previously been appointed by the Health of Towns Association Committee (of which Duncan was secretary) in 1844. The first case of cholera in Liverpool was officially recorded on 17 May 1832. Outside London, Liverpool's cholera caseload was the highest in England. Until the last case in September 1832, there were 4,977 reported instances of the disease (about 3% of the city's population), with 1,523 deaths (a mortality of 31%). Most of the affected, and particularly those who died, were poor, often cellar dwellers and frequently Irish immigrants. In a pamphlet, Duncan highlighted the problems of Liverpool that led to ill health, the 6,294 inhabited cellars with 20,168 inhabitants. Many of these cellars had no windows. He highlighted the narrowness of streets and that cellar inhabitants were without any place to deposit their "refuse matter" – in many instances, the inhabitants of the front houses in the "courts" and cellars made use of the conveniences of the courts so that the ashpits "become full to overflowing". While much had been done to improve the provision of drains and sewers, he pointed out that these had been principally main sewers and only very few of the side streets were so served. The 25 miles or so of sewers were unevenly distributed so that of the 20 miles of streets occupied by the "working population" only about four miles were sewered. Thomas Fresh, as the Inspector of Nuisances appointed by the council, is seen as the first EHP worked alongside Duncan.

When thinking about the reformers and key people in the 19th-century public any EHO/EHP should not ignore Fresh, as the first Inspector of Nuisances appointed under the private Sanatory Act of 1846. He had held an informal position prior to that. He came from a wealthy Lake District family and moved to Liverpool initially as an iron-founder and then policeman. He worked closely with Dr Duncan and reported to the Health Committee on cellar habitations, overcrowding, inadequate water supplies and sewage pollution of dwellings, accumulations of offensive matter, slaughterhouses, and nuisances. Fresh was appointed to enforce sanitary regulation and produced a quadrennial report on the scope of his work. However, despite council officers being expressly forbidden from all private businesses, Fresh invested his family inheritance in the development of 150 acres of land in what these days is known as Freshfield, between Liverpool and Southport. He was also involved in the formation of a Building Society. When this became known to the Council, he felt obliged to resign his position in 1859.[32] Thus, it can be said that the first EHP also had ethical questions to answer. He was not the last practitioner who had interests outside of their employment that could compromise their professional judgement or raise ethical questions or at least be perceived as such.

A pioneer who met resistance from those in power was Dr John Simon who was the City of London's first MOH and first Chief Medical Officer for England. Again, he was an advocate of the links between poor housing and poor health, but his reports were seen as an exaggeration by those with a vested interest. His public health message even when included in legislation was often watered down by politicians and vested interests.[33] It has been a recurring theme that often public and environmental health gained traction only when there were problems that affected more than just the poor – think of The Great Stink of 1858.

In the 19th century, cesspools were seen as the best way to store human waste, and in London some communal public toilets discharged straight into the River Thames. Joseph Bazalgette was London's Chief Municipal Engineer to the Metropolitan Board of Works: he designed and oversaw the construction of a sewerage system for London. However, there was ongoing conflict with the likes of Chadwick who because of the miasma theory believed that engineers such as Bazalgette created overly expensive solutions in collaboration with irrational local government at the time. Resistance to Bazalgette also included those who thought that it was not the role of government to build public infrastructure.[34] It became more accepted when private schemes faced near bankruptcy.[35]

In terms of environmental health, doing the right things (even if for the wrong reasons) and not being popular with the powers that be, has a long tradition. Many of the reformers discussed here were also driven by personal moral imperatives. One might ask where are the 21st-century equivalents in the environmental health profession in the 21st century. A question that keeps coming to mind is why the public and politicians seem to prefer cure to prevention of ill health. That could be because it is easier to show that an individual has been healed or been made well. Prevention of ill health or injury is more difficult to prove at the individual level and in the main needs a population approach.

The development and reorganisation of local government and public health

Any historical perspective would be incomplete without considering how local government and the public and environmental health role of local government has developed from the "vestrys" or "parishes" responsible for administering the Poor Law (and workhouses) to local councils. Local government has developed in a way that is peculiarly British (English?), and it can be difficult to explain to students from other European countries. Indeed, it has been said that the "history of local government in the UK, then, can be described as one rooted in these two dichotomous traditions: the centralising fetish of the state – the veritable 'Norman Yoke' – bolted on to the decentralised chaos of the Anglo-Saxon heritage."[36]

The Municipal Corporations Act 1835 was all part of the reforms to address "rotten boroughs". This was followed by the Public Health Act 1848 establishing local boards of health in towns, to regulate among other things sewerage and to prevent the spread of diseases. In municipal boroughs, the corporation selected the board; in other urban areas, the ratepayers elected the boards.

The Public Health Acts of 1873 and 1875 established organisations to administer both the poor law, and public health and sanitation. Urban sanitary authorities were created from the local boards of health, and rural sanitary districts were created from the poor law unions, so it was all an ad hoc development. The Local Government Act 1888 was the first systematic attempt to impose a standardised system of local government in England and with the Local Government Act 1894 there were different types of local authority counties, county boroughs, rural and urban district councils, and London County Council. Although there were reviews and some changes of status and boundaries, the next major reform did not occur until 1972 which provided for metropolitan counties (abolished by the Local Government Act 1985), as well as the "shire" non-metropolitan counties, metropolitan boroughs in the metropolitan counties (the major urban areas), and district councils. in the non-metropolitan counties. The two-tier level of government has gradually been eroded as more "unitary" authorities have been created largely from the former non-metropolitan counties with the districts abolished or in some cases by grouping districts and division of the county into two or more unitary authorities.

Gorsky et al.[37] have suggested that the bacteriological discoveries of Louis Pasteur legitimised extensions of local public health duties, which by 1900 included drainage, sanitation, safe water supplies, street cleansing, disinfection, disease notification and isolation, food safety, smallpox vaccination, and institutional care principally for "lunatics" and infirm older people. After the First World War, the "homes fit for heroes", agenda involved health departments in slum clearance. In 1948, the establishment of the NHS meant, as Gorsky et al.[37] suggested, that the organisational model was a compromise; limited public health functions were retained by local authorities, each under a MOH, while their pre-existing hospital and clinic services were transferred to new NHS bodies. Public Health's location then remained stable until 1974 when NHS reorganisation removed all but environmental health duties to health authorities and terminated the MOH role. It has been argued that declining morbidity and mortality from infectious diseases meant that health concerns lost their importance within environmental and housing policy while the MOH had continued to focus on tackling communicable diseases and health service administration after 1948, and in general had not been prepared to court controversy.[37] That attitude may have prevailed before then given that Crook[38] refers to one MoH as saying earlier in the 20th century that inspectors who attempted "to revolutionize suddenly and violently the whole of the bad habits of the people and the faulty construction of ancient houses", were utterly wrong-headed and would invariably end up engaged in unnecessary, expensive, and potentially unsuccessful litigation.

Arguably local government's public health role may also have been eroded by an expanding welfare state. This in turn reduced the role of local government in health advocacy such as on smoking cessation and reducing tobacco consumption. Some of the 19th-century public (or environmental) health functions have also been transferred to other agencies or privatised – one has only to think of public water supplies and sewerage where provision is privatised, and regulation is down to central government agencies. It seems from recent history that there very limited ethical obligations have been placed on privatised utilities.

The Health and Social Care Act 2012 reconfigured the structure of English public health, moving significant powers to local government with the Department of Health arguing this was returning public health home to its pre-1974 location. This was not strictly true as Directors of Public Health[39] (DPH) were at the upper tier of local government (where there was two-tier local government). The statutory responsibilities of the DPH are expected to match the public health duties of their local authority and there is a duty to write an annual report on the health of the local population and the authority's duty is to publish it. This raises the question of what contribution EHPs are making to this report.

Public and environmental health has become fragmented and attempts to improve co-ordination to bring them together have not been helped by austerity, and the failure of politicians to understand the connections as well as the inadequacies of governmental structures to reflect the view of Cicero that *Salus populi suprema lex esto* (the welfare (health) of the people shall be the supreme law). Further reforms of local government are proposed, and if there are more combined authorities with powerful mayors such as Andy Burnham in Great Manchester, there might be an opportunity to address this lack of co-ordination. However, politicians always have an eye on the next election and may not always support public health projects where the benefits may be longer term and after the next election.

In conclusion, reorganisations that have occurred thus far appear more to be the consequence of political and economic concerns rather than having any underlying real health or ethical basis. There has certainly been no desire to see decisions made as close to communities as possible, with central governments increasingly centralising control.

Environmental health: qualification and its representative bodies

It is worth thinking at this juncture why Crook[33] questioned why in Victorian and Edwardian Britain (primarily England) local sanitary inspection and the politics and ethics that constituted the sanitary inspector as a governmental agent, have been so neglected and sanitary inspectors should have received relatively scant attention. It still seems that the role and work of the EHO or EHP today is given insufficient recognition and perhaps standing, and this remains a challenge for their professional bodies.

The first move towards the creation of an association was the publication of a letter in *The Sanitary Record* calling for the establishment of such a body in London like one formed in Birmingham some time before.[40] In June 1883 at a meeting in London it was agreed that The Association of Public Sanitary Inspectors be formed. In January 1884 and at the age of 85, Edwin Chadwick was invited to be President.

A Century of Progress[35] sets out many of the developments of the environmental profession and the professional body, including some of the challenges, given that until the 1960s it was not a "graduate profession". At the close of the 19th century among the concerns of the Association was that of qualification and training. The Sanitary Institute set the examination for sanitary inspectors which was not a statutory qualification, although over 90 authorities required it. The Association was not happy with the examination, as it required no proper training and had become "the last resort of social failures". This led to the formation in 1899 of the Sanitary Inspectors

Examination Board with its main purpose the examinations for persons wishing to be appointed as sanitary inspectors under the Public Health (London) Act 1891. Although the Board had representatives from a range of bodies including the Royal Institute of Public Health and the Incorporated Society of Medical Officers, the Association was not represented. In 1908, the Council resolved to establish its own examination system and examination board, and successful completion of the two-part examination system led to the award of a diploma. Qualification and training always seemed to be a problem, with at one time the Sanitary Inspectors Examination Board for London appointments and the Royal Sanitary Institute (which eventually became the Royal Society for Public Health) for the provinces. In 1926, a joint board was established and the board for London appointments wound up.

As an aside in 1900 Johnson[35] refers to a libel action brought by the Association against a P. Anderson Graham who it was alleged had charged the whole body of sanitary inspectors with being corrupt and open to bribery so as to neglect their duties and connive with people on the adulteration of milk. A full retraction of the accusations was made before trial, but this demonstrates the importance of regulation of the profession and how ethics was and remains an important issue for EHPs.

After the Second World War, a new examining body was set up, the Public Health Inspectors Education Board with a large representation from the Association, although in effect this was reconstituted joint board. In 1957, the association changed its name to The Association of Public Health Inspectors (APHI) (although the journal remained *The Sanitarian* until 1964). It was at this time that the coat of arms was revised but retaining the five-pointed star which in heraldry is a sign of health and the motto *Amicus humani generis*. This has some implications for the ethics of the profession. The statutory designation of "Public Health Inspector" remained until the reorganisation of the National Health Service and Local Government in 1974.

It should be noted that developments in Scotland took their own path. The Royal Environmental Health Institute of Scotland (REHIS) was established in 1983 following the merger of the Royal Sanitary Association of Scotland (founded 1875) and the Scottish Institute of Environmental Health (founded in 1973) and incorporated by Royal Charter in 2001. The relationship between REHIS and the CIEH has at times been a little fraught.[41] There had been reciprocal arrangements for the environmental health qualifications North and South of the border until the closure of the Environmental Health Registration Board, but the CIEH replacement did not meet the same conditions that had previously been acceptable to REHIS.

A matter that had been a concern of the then Association in England and Wales throughout the early decades of its existence (and perhaps is relevant again today) was the insecurity of tenure of sanitary inspectors. The Sanitary Officers (outside London) Order 1910 made a substantial advance towards securing reasonable security of tenure but not to the full satisfaction of the

Council of the Association. As the result of pressure and perhaps changes in society as the result of the First World War, an Order to confer security of tenure on medical officers of health and sanitary inspectors was drafted but not issued until after the end of the war. This remained the situation until 1974 and is discussed below. This raises the question whether there is adequate regulation and security of tenure in local authorities in the present day and what impact does this have on the "ethics" of the profession.

Reorganisation, renaming, and regulation

Local government and NHS reorganisations in the early 1970s were perhaps the most dramatic for the environmental health profession and indeed have distinct relevance to the consideration of professional ethics. The Public Health Officers Regulations 1959 and the statutory designations of both "Public Health Inspector" and "Medical Officer of Health" were repealed, and almost all references in legislation to named officers were replaced with the term "proper officer of the local authority". The APHI had always been concerned that public health inspectors, at least in theory, worked under the direction of the MOH. As Parkinson[42] commented there had been long-standing tensions between MOHs and sanitary inspectors on issues such as the practical interpretation of the legislation that required the sanitary inspector to work "under the general direction" of the MOH.

The MOH's role was replaced with the Community Physician employed by the Area Health Authority. However, almost all the functions previously carried out by the Chief Public Health Inspector remained with local authorities (at the district level). These authorities could however appoint any suitably qualified person as an officer. The Report of the Study Group on Local Authority Management Structures (the "Bains Report") had recommended that a "Chief Environmental Health Officer" be appointed by each local authority as its "proper officer" for all environmental health functions, but this did not happen. The Ministry and the local authority associations supported the retention of the Public Health Inspectors Education and Registration Boards, although again the names would change to reflect the name of the profession.

In 1970, the APHI had published a Green Paper entitled "The Future of the Profession and the Association".[37] It was following this and local government reorganisation that the APHI changed its name yet again to The Environmental Health Officers Association and a code of Professional Conduct to apply to members was adopted in 1977 for the first time. It could be said that these changes completed the professionalisation of practitioners or at least their professional body as the General Council of the Association was empowered to remove from membership anyone contravening the code. The operation of the current code will be discussed later in this book.

At this time the Environmental Health Officers Education Board also changed its constitution and widened its powers so that it could control training, hold examinations, and award qualifications. At the same time the Public Health Inspectors Registration Board changed its name to reflect the change in name of the profession to Environmental Health Registration Board (EHRB). The Board registered students for the Diploma and awarded it, as well as to those whose training the EHRB considered was the equivalent of the Diploma. The Association also took over the administration of the Boards that had been responsible for overseeing the "modern" Diploma in place since 1964.

In the 1960s, the notion of a university degree as a route to qualification was also mooted and the first degrees were awarded in 1968 at Aston University and in 1967 Salford University started an environmental health degree. Since then, the number of universities offering an environmental health degree have varied.[43] Pressures on local authorities and universities have implications for the future of the profession. The viability of courses depends on the number of students, and if the numbers are too low, a course will be closed and this then leads to fewer new fully qualified EHOs and EHPs. A problem which is exacerbated by the reluctance of many local authorities to provide the necessary support for practical training for graduates. This is perhaps why it is possible these days to be registered with the CIEH as a food safety practitioner and housing health practitioner. The highest grade of professional achievement is Chartered Status recognising an individual's breadth of knowledge and experience and this person must be a member of the Institute.[44] There are, however, just under 2,000 of these out of a total membership of over 7,500.[45]

In 1981, the Association changed its name once again, this time to the "Institution of Environmental Health Officers" (IEHO) and commenced negotiations with the Privy Council for the grant of a Royal Charter which was granted in 1984. This change in 1981 included transferring to the Institution all the functions of the Education Board (more than just the administration) the thinking being that control of education and certification should be a recognisable characteristic of a profession. The EHRB, however, remained a separate entity. This remained the case until September 2021, so EHPs and related professionals who qualified in England, Wales, and Northern Ireland were registered with the EHRB. The EHRB formally closed in December 2021, although this remains a somewhat contentious issue for some given the professional body now had responsibility for registration. The financial challenges faced by the professional body led to a policy of "widening the membership" which raised questions as to whether it was now sufficiently impartial, as the motion at the 2023 Annual General Meeting made clear.[46] A criticism of the EHRB was that admittance was for life without the need to determine that the person had maintained their competencies or professional ethics.

In May 2020, during the COVID pandemic when asked who was considered competent to perform the role of Tier-2 Clinical Contact Caseworker the response was to release the EHRB register to assist EHPs wishing to apply to be recruited by NHS professionals. So, it would appear the Registration Board had value (not denied by the CIEH), but the proposers of this successful motion could not understand why the Board of Trustees allowed the EHRB to be wound up, when it was the conditions of registration that needed to be reviewed, not the EHRB itself. Indeed, it was the debates at this and subsequent General Meetings that prompted the idea of this book. The CIEH has suggested that the EHRB register was released during the pandemic because it was the most comprehensive list of registered EHPs at that time, even though some on the register had committed serious crimes, were dead, or had not worked in environmental health for decades.[47]

In 1994, the IEHO was given consent for a change of name becoming the "Chartered Institute of Environmental Health (CIEH)". The Royal Charter has been amended on several occasions most recently in 2016. The objects for which the Chartered Institute is established are to promote for the public benefit the theory and science of environmental health in all its aspects and to disseminate knowledge about environmental health. On professional regulation the Charter gives the CIEH the power to promote and maintain high standards of practice and professional conduct among its members and those others registered in the field of environmental health. It also has the power to create Registers of Chartered Practitioners and of Non-Chartered Practitioners with such sub-sections in such areas of environmental health as the Chartered Institute may determine. This gives the CIEH considerable powers over its members and those who are on the new register, but there is no requirement to be a member to practise as an EHO/EHP if registered by the EHRB.

In addition, EHOs/EHPs have moved from being a wholly local government profession (with the occasional representative in central government or devolved administrations) to one where there are many EHPs in private practice or employed in commercial or industrial businesses. Thus, members of the profession are more than "enforcers" of environmental health laws. This in part also explains why the term "practitioner" is used by the CIEH rather than "officer" (used in Scotland) as the latter implies the role is as a regulator or official, when that is no longer the role of a substantial number of people with environmental health qualifications. These changes have been significant in the past 50 plus years – as exemplified by one author here who qualified as a public health inspector but now has the status of a chartered EHP but has not been employed by a local authority since 1980. This though has implications for the regulation of the environmental health profession.

Environmental health and regulated professions in the UK

Environmental health is one of the professions in the UK that is regulated, but the situation is different in Scotland from the rest of the UK. The term "regulated profession" is defined in the Professional Qualifications Act 2022 (which revoked the EU-based system for recognising professional qualifications) and establishes a new approach based on regulator autonomy. It defines a profession which is regulated by law in the UK or a part of the UK. "Profession" is defined as to include an occupation or trade, or any subdivision and specialism within a profession. A profession is regulated by law where there is a legal requirement to have certain qualifications or experience (or meet an alternative condition or requirement) to undertake certain professional activities or use a protected title. The "Regulator" is also a defined term in the Professional Qualifications Act and means a person with functions under legislation relating to the regulation of a regulated profession.

Regulators carry out a range of functions in relation to the professions they regulate, including making sure individuals have the necessary qualifications and/or experience to practise the profession and taking any necessary enforcement action. Regulators include bodies to which regulatory functions have been delegated, where the delegation is authorised by legislation and irrespective of whether the regulator is specified in the legislation. So far as EHPs or EHOs are concerned it is different between Scotland and the rest of the UK.

Other than Scotland the regulatory authority is the CIEH (Chartered Body), and the regulated profession is that of EHP (EHO in Scotland) whether registered or chartered. The regulation type generally is accreditation (see later) but is "certification" for non-chartered EHPs. Certification means there are no restrictions on who can carry out activities, but by law only certified professionals can use the professional title. This raises the question as to whether this provides adequate protection for the public health and over the ethical standards of those undertaking environmental health activities. According to the government website, the protected titles are Chartered Environmental Health Practitioner[48] and Registered Environmental Health Practitioner (REnvH) and there are no reserved activities, so restrictions apply only to the use of title.[49,50] In a hearing under the code it was indicated that while the titles of EHO or environmental health professional "did not have protected title status in this jurisdiction, the chartered title did in accordance with the CIEH's charter". This has some bearing on how the code of ethics can be applied (but does not ensure that those undertaking environmental health work adhere to the code of ethics and standards of behaviour).

The route to qualification is an accredited BSc, BSc apprenticeship, or MSc followed by a period of work-based activity which is then assessed by CIEH. Successful completion of all these elements leads to registration as an EHP. As we have seen above, since 2022 the CIEH has had ownership of the

EHRB Registers and established its own Registers of Environmental Health Practitioners, Chartered Environmental Health Practitioners, Food Safety Practitioners, and Housing and Health Practitioners.

The Public Health (Scotland) Act 2008 requires local authorities there to designate competent persons under that Act. Such persons must meet the prescribed requirements of the REHIS and be eligible to hold the title EHO. This applies to the protected titles under the Royal Charter of Environmental Health Officer (EHO), Chartered Environmental Health Officer (Ch.EHO MREHIS), and Chartered Environmental Health Officer (Ch.EHO FREHIS).

The Public Health (Scotland) Act 2008 Designation of Competent Persons Regulations 2009 (SI 2009 No 301) requires that to be eligible for designation as a local authority competent person, that person must be an EHO having completed an accredited BSc (Hons) degree, plus the associated practical training and professional examination including portfolio and professional interview, and have a minimum of two years' experience working as an EHO at a local authority or equivalent. This does not stipulate any number to be employed, however. Thus, local authorities in Scotland expect all professionals holding the title EHO to meet the requirements of the Institute.

Under the Professional Qualifications Act 2022 Act, REHIS is the regulatory authority, and the regulation type is "accreditation" (as it is for Chartered EHPs in other parts of the UK) which means in theory there are no restrictions on who can carry out activities or use titles, but accreditation may demonstrate that individuals have achieved certain qualifications or experience but as noted above the titles are protected by the Royal Charter.

Conclusion

The work of EHPs and society has changed enormously since the days of Fresh and Duncan, but the notion of health equity should remain fundamental to the philosophy and ethos of the EHPs. Ian Watts[51] has in the context of the economic issues facing Britain referred to the American moral, legal, and political philosopher John Rawls and the notion of reciprocity. That is, all who take part in a well-ordered society are equally deserving of the benefits of cooperation. Gains to the more advantaged should not be made at the expense of the less advantaged. Watts suggests there is a distinct lack of reciprocity in a neoliberal Britain. An illustration of this that will resonate with many EHPs goes back to the COVID pandemic years when the middle classes in houses with gardens became leisured and more able to cope with lockdowns leaving the less well-off cooped up in small flats or houses in multiple occupation and/or working to deliver essential services.

Environmental health is a component of public health, and it is apparent that improvements to public health have come about as the result of people who did not court popularity but were driven by other motives. So perhaps this is something that should be noted by EHPs/EHOs individually and their professional bodies collectively.

Much of the history of the CIEH and the profession has been about improving its position in society, and it is not clear that until relatively recently this had any ethical dimension, at least not explicitly. Even now with changes to recruitment and training of those undertaking environmental health work in local government, one must ask whether ethics or any ethical code plays any part in how they perform their duties.

It might be that there is a reliance on the general standards of behaviour expected of local government officers and enforced by the employer and the Seven Principles of Public Life which outline the ethical standards to which those working in the public sector are expected to adhere (the Nolan Principles). These apply to all people appointed to work in local government. However, the increasing number of EHPs employed outside local government including in private practice are beyond the scope of these principles.

How the regulation of the environmental health profession in particular works in practice and ensures that ethical matters are given sufficient consideration is considered in the next chapter.

Notes and References

1 See https://public-domain-poetry.com/alexander-pope/epilogue-to-the-satires-in-two-dialogues-dialogue-ii-28823
2 Coggon J and Viems AM (2017) *Public Health Ethics in Practice – A Background Paper on Public Health Ethics for the UK Public Health Skills and Knowledge Framework*, Crown Copyright.
3 https://www.fph.org.uk/public-health-ethics-and-law
4 Dhesi S, (2019) *Tackling Health Inequalities – Reinventing the Role of Environmental Health, Routledge Focus on Environmental Health Series*, Routledge, Abingdon, Oxon.
5 Traditionally solicitors could only represent clients in the Magistrates' Court but can qualify to obtain higher rights of audience meaning like barristers, they can represent clients in the Crown Court and appeal courts, these are Higher Court Advocates. All solicitors, you must apply to be registered with the Solicitors Regulation Authority. This register is known as the roll of solicitors in England and Wales and entitles a person to practise as a solicitor and is necessary before being able to join the Law Society which is the professional body for solicitors in England and Wales.
6 Registered Environmental Health Practitioner – is also a protected title which means only those people who are registered with the CIEH can use the title but such a person does not have to be a member.
7 A private communication from the CIEH (June 2025) indicated that there is no data on the proportion of those completing an accredited course who do not register. There has never been 100% registration as some EH graduates pursue careers in other industries or overseas – "or choose to work in environmental health without registration". Anecdotally, a lack of placements to complete the necessary practical experience or lack of support from employers for candidates on placements seems to be a "significant issue" as to why not all those who graduate can or do register.
8 https://www.cieh.org/media/5249/cieh-workforce-survey-report-for-england.pdf

9 Raworth K, (2017) *Doughnut Economics - Seven Ways to Think Like a 21st Century Economist.* Penguin Random House, London.

10 These two terms are used because different professional bodies use the two terms, for example, in the UK the regulated profession is EHP except in Scotland where it is EHO.

11 See, for example, Saks M, (2012) Defining a Profession: The Role of Knowledge and Expertise. *Professions & Professionalism*, Vol. 2 No. 1, pp. 1–10 (updated 2016). https://doi.org/10.7577/pp.v2i1.151

12 Australian Council of Professions (2003) What Is a Profession, at https://professions.org.au

13 Veatch RM, (2009) The Art of Medicine, The Sources of Professional Ethics: Why Professions Fail. *The Lancet*, Vol. 373 (March 21), pp. 1000–1001.

14 Flick C, (1980) The Movement for Smoke Abatement in 19th-Century Britain. *Technology and Culture*, Vol. 21, No. 1 pp. 29–50. The Johns Hopkins University Press and the Society for the History of Technology. https://www.jstor.org/stable/3103986

15 See Battersby SA, (2022) Historical Context, Philosophy, and Principles of Environmental Health, in Battersby SA Ed., *Clay's Handbook of Environmental Health.* 22nd Edn, pp. 1–28, Routledge.

16 See Parkinson N, (2014) The Health of Towns Association and the Genesis of the Environmental Health Practitioner. *Journal of Environmental Health Research*, Vol. 14, No. 1, pp. 5–16.

17 See Parkinson N, (2013) Thomas Fresh (1803–1861), Inspector of Nuisances, Liverpool's First Public Health Officer. *Journal of Medical Biography*, Vol. 21, No. 4, (November), pp. 238–249. https://doi.org/10.1177/0967772013479277. Epub 2013 Jul 26.

18 See also https://www.cieh.org/digital-archive/history-of-environmental-health/

19 https://www.britannica.com/summary/The-Enlightenment-Key-Facts

20 Institute of Medicine (US) (1988) *Committee for the Study of the Future of Public Health. The Future of Public Health.* National Academies Press (US), Washington, DC; 3, A History of the Public Health System. Available from: https://www.ncbi.nlm.nih.gov/books/NBK218224/

21 See https://www.worldhistory.org/article/2221/why-the-industrial-revolution-started-in-britain/ and https://www.itakehistory.com/post/trading-empires-england-vs-france-in-the-proto-industrial-age, for example

22 https://www.britannica.com/topic/Christian-Socialism

23 Driver J, (2022) The History of Utilitarianism. *The Stanford Encyclopedia of Philosophy* (Winter 2022 Edition), Edward N. Zalta and Uri Nodelman (eds.), https://plato.stanford.edu/archives/win2022/entries/utilitarianism-history.

24 In 1832, the government appointed a royal commission to investigate the workings of the Poor Law, and in 1833, Chadwick was promoted to Chief Commissioner and recommended more centralisation of the Poor Law which also needed to be reformed so as to deter people from making unnecessary demands on public funds. Centralisation was not popular at the House of Commons which did not want to pay responsible state officials a salary from central funds if local men, regardless of their capabilities, could be paid a pittance from the Poor Rates. See more at: https://web.archive.org/web/20170202032848/http://www.lshtm.ac.uk/library/archives/history/frieze/chadwick.html

25 Vaishali MP, Boopathy N, (2024) Edwin Chadwick: A Pioneer of Public Health Reform and His Role in Sanitary Awakening. *Cureus*, Vol. 16, No. 9 (September 7), p. e68858. https://doi.org/10.7759/cureus.68858. PMID: 39376875; PMCID: PMC11457289.

26 Hamlin C, (1997) *Public Health and Social Justice in the Age of Chadwick: Britain, 1800–1854.* Cambridge University Press, Cambridge.

27 Stewart J, Ed. (2017) *Pioneers in Public Health*, Routledge Focus on Environmental Health Series, Routledge, Abingdon, Oxon.
28 Ringen K, (1979) Edwin Chadwick, the Market Ideology, and Sanitary Reform: On the Nature of the 19th-Century Public Health Movement. *International Journal of Health Services*, Vol. 9, No. 1, pp. 107–120. https://doi.org/10.2190/LR4G-X2NK-9363-F1EC. PMID: 370032 (Élie Halévy (1870–1937) was a French historian, author of a detailed general account of 19th-century British history and was right wing. Eric Hobsbawm (1917–2012) was a Marxist historian who wrote among other things about the rise of industrial capitalism).
29 Mooney G, (2000) Review of Public Health and Social Justice in the Age of Chadwick: Britain, 1800–1854, and: The Gospel of Germs: Men, Women, and the Microbe in American Life. *Journal of Health Politics, Policy and Law*, Vol. 25, No. 4, pp. 775–781.*Project MUSE*, https://muse.jhu.edu/article/15490.
30 Hamlin C, (1996) Edwin Chadwick, "Mutton Medicine," and the Fever Question. *Bulletin of the History of Medicine*, Vol. 70, No. 2 (Summer), pp. 233–265.
31 Battersby SA, (2018) Dr William Duncan, in Stewart J, Ed, *Pioneers in Public Health: Lessons from History*., Routledge, Abingdon Oxon. pp.56–65.
32 Parkinson N, (2018) Thomas Fresh, in Stewart J, Ed, *Pioneers in Public Health: Lessons from History*, Routledge. Abingdon, Oxon, pp.11–23.
33 Page A, (2018) Sir John Simon, in Stewart J, Ed, *Pioneers in Public Health: Lessons from History*, Routledge. Abingdon, Oxon. pp.24–32.
34 Page A, (2018) Sir Joseph Bazelegette, in Stewart J, Ed, *Pioneers in Public Health: Lessons from History*, Routledge. Abingdon, Oxon. pp.40–45.
35 It should not be forgotten that another vested interest in dealing with human waste was the international trade in guano from South America used by farmers as fertiliser in the fields. See Kinsley LJ, (2019) *Guano and British Victorians: An Environmental History of a Commodity of Nature*. Dissertation submitted to the University of Bristol in accordance with the requirements of the degree of Doctor of Philosophy in the Faculty of Arts, School of Humanities, University of Bristol. Available at https://research-information.bris.ac.uk/ws/portalfiles/portal/231791845/LKinsleyPhDThesisMarch2020.pdf
36 The Warwick Commission (2012) Elected Mayors and City leadership, Summary Report of the Third Warwick Commission, Warwick University. https://warwick.ac.uk/research/warwickcommission/electedmayors/summaryreport/
37 Gorsky M, Lock K, Hogarth S, (2014) Public Health and English Local Government: Historical Perspectives on the Impact of 'Returning Home'. *Journal of Public Health*, Vol. 36, No. 4, (December), pp. 546–551. https://doi.org/10.1093/pubmed/fdt131. Epub 2014 Jan 27. PMID: 24472776; PMCID: PMC4245898.
38 Crook T, (2007) Sanitary Inspection and the Public Sphere in Late Victorian and Edwardian Britain: A Case Study in Liberal Governance. *Social History*, Vol. 32, No. 4, pp. 369–393. https://doi.org/10.1080/03071020701616654 quoting Blyth A W., 'The Education, Status and Emoluments of Sanitary Inspectors', *Journal of the Sanitary Institute*, XVIII (1897), 197–8. *cited by* Rutherford D, (1912) Conference of Sanitary Inspectors: Presidential Address. *Journal of the Royal Sanitary Institute*, Vol. XXXIII, p. 76.
39 The Association of Directors of Public Health has existed in its current form since 1989, but there has been an equivalent membership body for more than 160 years when in 1856 when the Metropolitan Association of Medical Officers was formed. The Faculty of Public Health is the standard setting body for public health specialists within the UK, setting standards for training, examination, and specialist practice across the four countries of the UK.
40 Johnson R, (1983) *A Century of Progress - The history of the Institution of Environmental Health Officers 1883–1983*, CIEH, London.

41 This arose primarily because the CIEH established a branch in Scotland and also claimed they represented environmental health for the whole UK.

42 Parkinson N, (2015) 'Inspector of Nuisances' to 'Environmental Health Practitioner': A Case Study of Title Change in the Professionalisation Process. *Journal of Environmental Health Research*, Vol 15, No. 1 (December), pp. 4–22.

43 Courses accredited by the CIEH can be found at https://www.cieh.org/what-is-environmental-health/how-do-i-become-an-ehp/bsc-enviromental-health-degree/#four. There is also BSc EHP apprenticeship programme. A number of universities offer a post-graduate MSc course; see https://www.cieh.org/what-is-environmental-health/how-do-i-become-an-ehp/postgraduate-msc/

44 They must have an environmental health degree, be registered, with a minimum of three years' experience of environmental health work and three years of compliant CPD records.

45 CIEH Annual Report 2024, https://www.cieh.org/media/ivsjq2sp/cieh-annual-report-2024.pdf

46 https://www.cieh.org/media/0eqpuwce/minutes-of-agm-12-september-2023.pdf

47 Personal communication from Jon Buttolph 1 May 2025.

48 https://www.regulated-professions.service.gov.uk/professions/chartered-environmental-health-practitioner-1

49 https://www.regulated-professions.service.gov.uk/professions/chartered-environmental-health-practitioner-1

50 Strangely and unlike for Scotland there are two separate pages on the government website for registered and chartered EHPs. In Scotland, EHO and Chartered EHO (including Fellows of REHIS) are all protected titles listed on the same page.

51 Watts I, (2025) Could John Rawls Save the Labour Party? *New Statesman*, 30 January.

2 "Ethics" and relevance to EHPs

Introduction

Ethics is the principle of moral conduct that makes a distinction between good and bad/ evil, right and wrong, virtue and non-virtue. The word ethics is derived from a Greek word "ethos" meaning character. So, as a general point it can be considered a branch of knowledge that governs right and wrong conducts and behaviours of an individual, profession, group, or organisation. There are both general ethics and professional ethics and both involve principles of good or moral behaviour. Ethics in general (global ethics) deals with moral values and the responsibilities of humans as individuals encompassing such matters as human rights, social justice, and environmental sustainability. In general, professional ethics are specific to the profession and focus on standards of conduct and behaviour within that specific field. From that we can see that in the environmental health sphere there is a clear overlap and considered later in this chapter.

While ethics can also refer to the philosophical study of morality, known as moral philosophy, this is not the direct concern of this book. That said ethics, and morality are now used almost interchangeably in many contexts. There was a brief section in the 22nd edition of *Clay's Handbook of Environmental Health*[1] on ethics, and this book develops in more depth what is included there.

Perhaps also fundamental to the consideration of ethics for a "health and care" profession (the sector in which the environmental health profession is entered on the regulated professions register in Britain)[2] is to define "health". This has an important bearing on the work of the environmental health officer (or practitioner) (EHO/EHP) but also must be reflected in the professional ethics of EHPs/EHOs (and be relevant when considering their professional codes of ethics). As we have seen from the history in the previous chapter, there has been some difference between the "medical" model of health and that applied by environmental health professionals. For clinicians, health often does just mean the absence of disease.

Medical ethics are also not of much assistance here. It has been argued that public health ethics should take account of the population approach.

DOI: 10.1201/9781003625339-2

That health is shared across society (not just individuals in isolation) and to think in terms of social co-ordination incorporating measures that target whole populations[3] or arguably whole sections of the population. In some respects, EHOs/EHPs and in practice the environmental health profession does both.

It is beyond the scope of this book to go much further and define "environmental health".[4] This is not least because it is a term that can have different interpretations in different countries, as was apparent from discussions the authors have had with the European Office of the World Health Organisation (WHO). It is also addressed in more detail in *Clay's*.[1] Suffice to say, unlike medicine which tends to operate in a silo and be more narrowly focussed, environmental health should take a more holistic approach. Hence, the claim sometimes made is that they are "on the ground problem solvers" – only possible for fully qualified EHPs/EHOs.

It has been said that[5] ethics is a set of values that guide the decisions of EHPs, influence actions, and give purpose to the lives of EHPs. Professional and personal ethics in environmental health practice are the cornerstones that help good decisions to be made which protect public health and promote the public's trust in the profession. Ethics not only guide "our decisions, but they also help us navigate complex situations that have no easy or clear answers".[5]

This chapter will develop an analysis of professional ethics in the context of EHPs looking also at "ethical" standards of some EHP professional bodies around the world and other professional bodies. It will consider the impact of statutory regulation of the profession. Professional ethics must be based on serving the interests of the public to protect or improve public and environmental health – that is reducing the environmental stressors that impact on individual and public health. This chapter will explore how this affects EHPs whether in local government or in private practice when the profession is not governed or protected by statute. This chapter will also examine how the code of ethics of the Chartered Institute of Environmental Health (CIEH) and other professional bodies including environmental health, work, and how they are enforced or implemented. Is there an argument that being a statutory appointment or at least having a similar situation as Scotland in the rest of the UK would aid professional ethics? Do financial considerations – maximising membership fees compromise regulation of a profession and the ethics of the regulator of the profession, and should there be a more independent registration system?[6] It examines some of the ethical questions; for example, with "scores on the doors" on food hygiene ratings and charges for licensing of housing in the context of "professional ethics", although this also arises again in the next chapter when discussing law and ethics.

This chapter considers what has proved to be helpful in developing status and influence of other professions. Without being "defensive" membership of a profession, with a well-defined ethical code of practice and rigorous criteria for inclusion, and which is properly regulated and enforced, has great benefits such that there should be:

- Reputational status of individuals – there is an explicit statement of individual expectations; knowledge; technical expertise; skill set; and trustworthiness. Individuals also share in the collective reputation.
- Increased public confidence in an individual's professional judgement.
- Collective support of fellow members against litigation and pressure to comply with inappropriate requirements of employers.
- Enhanced employment prospects – membership should be a benchmark for employment, defining roles for which only members of the profession are eligible for appointment.
- Access to, and a requirement for personal professional development and to the development of the profession.

Definition of health

Consideration of what is meant by health has a bearing on their approach to work and ethics for an EHP/EHO. The definition of health often used by EHPs and EHOs is that included in the preamble to the constitution of the WHO:[7]

> *Health is a state of complete physical, mental and social well-being and not merely the absence of disease or infirmity*

The principles include that the enjoyment of the highest attainable standard of health is one of the fundamental rights of every human being without distinction of race, religion, political belief, economic or social condition. Additionally, an important implication of this definition is that mental health is more than just the absence of mental disorders or disabilities.

This definition has been taken for granted and accepted as a given by EHOs/EHPs, but it has come in for criticism in recent years, particularly from some politicians and others who believe there is an element of "overreach". It is arguable that this is based on the interpretation or misinterpretation of the word "complete". Such criticisms are answered by Schrammer[8] who explains why this is a misinterpretation and this is explained below.

One criticism that has been made is that the linking of health and well-being is a positive ideal that has given rise to problems such as a tendency to define all social problems from war to street crime as "health" problems. This does raise the issue of whether EHOs/EHPs can be guilty of this? It is perhaps not unreasonable given that this profession has from the very beginning dealt with "nuisances" which may not be thought or recognised as always posing a risk to health in the medical sense (although in the 19th century there was a good chance they did). This is particularly so as the statutory nuisance provisions now distinguish "nuisance" and "prejudicial to health" as two separate limbs. Case Law[9] (having looked at the long history of the law – so that "the reach of this statute to disease and ill-health is too long-standing and deep-rooted to be susceptible now to any different interpretation") has determined that in this context the threat of physical injury does not fall within the scope of "prejudicial to health" even if defined as "injurious or likely to cause injury to health" – whereas in the real world falling down dangerous and steep stairs and breaking limbs or worse might be seen as a health problem and would justify medical attention.

Schrammer[8] has suggested quite persuasively that the World Health Organization (WHO) defines health as holistic health, not as perfect health. After looking at the development of the constitution the word "complete" was inserted at the very final stages of drafting and was an editorial amendment, rather than a substantial change, which would have had extensive debate that did not happen. In essence, the word complete jointly describes the three aspects of wellbeing and perhaps could be replaced by "total" or "overall" or even "comprehensive". It is suggested that all the word does is add a "rhetorical contrast to the second part of the sentence that denies the sufficiency of the absence of disease or infirmity for health". It might have been better Schrammer suggests, to express health as a "state of complete well-being, that is, a state that comprises physical, mental and social elements".

In terms of public and environmental health perspective, it is obvious, despite the legal interpretation of what is "prejudicial to health",[8] that health is "more than the absence of disease". Schrammer talks of an "holistic" health in the context of the WHO definition rather than perfect health and given the environmental stressors that EHOs/EHPs know impact health, that seems self-evident. It should therefore be seen that health is a better condition than "medical normality" – the absence of disease (or unintentional injury). People who live in poverty and poor housing conditions might not obviously suffer from any disease caused directly by those conditions but can be living such that they cannot maintain minimal health as defined by WHO. It should be noted that the Housing Act 2004 defines health as including mental health, which means for the first time housing law in England and Wales has a definition of health close to that of the WHO. This does also mean that EHPs need to understand what that means – and in the context of housing that threats to health and safety from housing deficiencies must be such that the outcomes from an occurrence could justify medical intervention.

The Schrammer approach allows for comparison of different groups to be able to identify health and social inequity, something that should be of concern to all environmental health professionals. Health equity is therefore something that should be reflected in a code of ethics. Registration with the CIEH is available to a wider group than what traditionally might have been seen as an EHO or EHP. Now there are registered food safety practitioners and registered housing practitioners with various routes to registration, but which by definition are not able to take a holistic approach to health or recognise the issue of health equity.

In the context of considering the definition of health and its relevance to ethics, it is useful to consider two related terms. Health equality is about giving everyone the same opportunities, care, and services. An EHO/EHP might deal with every food premises the same without regard to all the risk factors or provide the same information to everyone believing that, as long as they treat everyone the same, they are not behaving in a biased way.

Health equity however means identifying and stopping institutional and discriminatory barriers that lead to health inequities and inequality. While this can include factors such as racism and sexism it also includes factors such as poverty and unequal distribution of resources. Health equity takes account of different cultures, access to resources, and socio-economic status. For example, poverty can affect health by reducing access to nutritious food, or warm dry housing, increasing stress and forcing people to live in unsafe and unhealthy circumstances. In a health equity framework, health professionals consider the impact of the social and strategies for addressing them.

As an aside, and perhaps also relevant to ethics, is Einstein's often used quote "The environment is everything that isn't me".[10] It suggests that everything around us and all life on the planet as well as human-made structures and possibly societal constructs, all form our environment. It highlights the importance of what is around us. It is apparent that this is more than just wildlife and natural ecosystems. This leads us to think of the environment as all things beyond our own individual being. Yet we as individuals and society through our actions can have an impact on the environment – some impacts may be good for some, but bad for other human beings.

What do we mean by professional ethics?

At its simplest a profession can also be defined as any type of work requiring special training or a particular skill and involves a high level of education, but this chapter will delve deeper. In the context of environmental health back in 1958, Chanlett set out some detailed thoughts as to what makes a profession.[11] These can be summarised as:

1 It must satisfy an indispensable social need and be based upon well-established and socially accepted scientific principles.
2 Require adequate pre-professional and cultural training.
3 Require possession of a body of specialised and systematised knowledge.
4 Have skills which the public does not possess.
5 Have developed techniques as the result of tested experience.
6 Require the exercise of discretion and judgement in the performance of duty contrasting with work subject to immediate direction and supervision.
7 Its work must be beneficial, but not subject to standardisation or control in terms of units of performance or time.

8 Have a group consciousness designed to extend scientific knowledge in technical language.
9 Have sufficient self-impelling power to retain its members throughout life and not be a mere steppingstone to other occupations.
10 Recognise its obligations to society by insisting that its members live up to an established code of ethics.

Perhaps it can be summarised by the key aspects of what defines a profession as a combination of the following attributes:

- a degree of licence to practice – typically with some form of statutory underpinning
- a body of knowledge or expertise
- a code of conduct which is binding on the membership.

Qualified EHPs today would seem to fit the criteria set out by Chanlett, perhaps some more than others but regrettably and in the typical shortsightedness, of England in particular, not all employers wish to appoint fully qualified EHPs (with their transferrable skills) to carry out what historically have been functions of the EHO/EHP. Inevitably we discuss the notion of some statutory underpinning of environmental health work.

Employing the spirit of openness and recognising "obligations to society" it is our view that the ethical code for any profession should be readily and publicly available. When any professional enters the work force, their duties involve decision-making, and EHOs and EHPs[12] are no different but they also come into close contact with the public. The nature of the work indicates that EHOs or EHPs should operate to high ethical standards. That is not true for others who may undertake tasks that in the past EHOs/EHPs have undertaken and thus the service provided could be less than optimal. It should be seen as a positive and essential to employ personnel with a strong ethical basis.

The role of an EHP involves checking compliance with legal standards and acting to protect and even improve public health and safety (not always the same). How they ensure compliance and resolve problems and secure improved public health requires a range of skills. Issues can and do arise when the law or public administrative structures and management are inadequate or non-existent for the environmental health problem when the policy or practise of employers is inimical to public health. Recent examples include the issue of smoking and vaping as examples or more recently the COVID pandemic or back to when AIDS/HIV was first identified. Vaping and e-cigarettes are supposedly less harmful than the use of tobacco but are not without health effects.[13] This can pose an ethical dilemma; what to do or what can be done in such circumstances. For instance, what should the approach be when it is local practise not to use powers in law that are intended to protect the health and safety of occupiers? What if the legislation fails to keep up with societal and public health needs?

For those working in local councils (which came into being largely to improve and protect public health), there can be the perceived limitation of the doctrine of *ultra vires*[14] despite legislative attempts to reduce its impact in the UK, there still can be a fear of acting outside a council's powers when addressing a real public or environmental health problem. Sometimes often forgotten is that local councils are public authorities (and part of the state) that fall within the scope of the Human Rights Act 1998 and must act in accordance with the European Convention on Human Rights (ECHR). The following chapter looks in more detail at the law and ethics.

Traditionally, ethics referred to the philosophical study of morality, the latter being a more or less systematic set of beliefs, usually held in common by a group, about how people should live. The term "ethics" may not only refer to the philosophical study of the concepts of moral right and wrong and moral good and bad, but also to any philosophical theory of what is morally right and wrong or morally good and bad, as well as to any system or code of moral rules, principles, or values. The last may be associated with particular religions, cultures, professions, or virtually any other group that is at least partly characterised by its moral outlook.

Ethics has been said to be a system of moral principles that guide how people make decisions and live their lives. As Hugman suggests,[15] ethics should be the concern of everyone and certainly we cannot avoid facing moral issues and therefore making moral decisions. This is particularly true of the environmental health professional who every day encounters the public and whose role is to safeguard public health from stressors whether at home, work, or at leisure. While for many this means enforcing legal provisions, the problems they encounter may not fit neatly into the law as they see it or interpret it. Indeed as "street-level bureaucrats" (a concept first coined by Michael Lipsky in 1969)[16] who are responsible for policy implementation on the ground but who have considerable discretion in how that is done and when they are in daily contact with citizens. This imposes (or should impose) some moral responsibility, and it is arguable that this requires a professional who conducts themselves ethically. Employing professionals who are governed by a code of ethics or code of professional practice should in theory at least reduce the need for direct management or oversight. This includes the need to develop over prescriptive rules which can be rapidly overtaken by change of circumstance.

There is a question also whether environmental health is seen as a "vocation", or can it become so? This is a term that in the past has had religious connotations as being "a calling" but the OED[17] suggests that modern definitions include "a profession, way of life, course of action, etc., which a person feels that it is his or her duty or destiny to follow, or for which he or she feels particularly suited" and also "the means by which a person makes a living; a person's employment or main occupation, especially regarded as requiring dedication". A vocation also delivers a sense of doing something worthwhile in life and that should be true of the EHP/EHO.

It is often assumed that by the very nature of the work EHPs or EHOs must be ethical otherwise why would they want to safeguard public health and indeed intervene to reduce the impact of environmental stressors on the public and individuals, but it does not always follow. EHOs/EHPs do not have the equivalent of the Hippocratic oath which requires physicians to uphold specific ethical standards. These days the General Medical Council (GMC) has guidance on good medical practice[18] which includes updated professional standards with a focus on behaviours and values. Yet despite all this it cannot be said that all clinicians have acted ethically over the years. As already indicated public and environmental health ethics should draw from what has been called "political philosophy", rather than medical ethics[3] even if it is not possible for practitioners to examine problems in philosophical depth. Some frameworks and models are needed to assist activity and decision-making (including on policy). These are needed in addition to general concerns of professional ethics which as we will see are often more concerned with behaviours such as honesty and transparency.

The thinking that whatever an EHO or EHP is doing must be inherently good is perhaps a reflection of virtue ethics, one of the common ethical theories.[19] This ethical theory places emphasis on the character of the moral agent. It focuses on the development of the individual's character and understanding of the virtues that make up good character. This theory holds that a good and moral person will naturally act in the right way, and that this is the best way to determine right and wrong. Virtue ethics places importance on understanding the virtues, such as courage, justice, and wisdom, that make up a good character. Other approaches to ethics might emphasise duties or rules, being concerned with what people do, not with the consequences of their actions (deontology) or might emphasise the consequences of actions (consequentialism) of which utilitarianism is an example.

An example of virtue ethics is what is also called Aristotelian Ethics, which posits that good habits lead to a good character. Being honest, always truthful no matter of the difficulty or consequences leads to integrity, which then influences actions positively. One can argue that professionally that is what the EHP/EHO should have as a characteristic. Yet in their work there may be several "truths" in that there may be several options for addressing a problem, but which may be mutually contradictory. This would suggest that these be spelled out explicitly and that action to address the problem should not be dictated by a rigid approach. This also implies that the professional should take time to reflect on the best course of action for resolving the problem.

Kantian ethics (named after the German philosopher Immanuel Kant) is an example of deontology which in effect says that we should act in a way that we would want everyone to act. An example would be providing help and advice to secure resolution of a problem even when it might be beyond what you are required to do as part of your employment but is the logical and rational thing to do, but because you would expect such help in similar circumstances rather than just out of pity. Kant would suggest that what might appear to be a humanitarian act is not morally good if done out of pity.

Utilitarianism nowadays is representative of what is termed consequentialism because it is concerned with the outcome of actions. So a decision to invest in a project which will save many lives such as a vaccination programme as opposed to investing in a project that will make life better for a few, would be a utilitarian decision because it creates the greatest welfare (health and happiness) for the largest number of people.

Thinking further in terms of environmental health, suppose someone needs help because they are renting poor conditions that are hazardous to health. It can be said that in modern ethics a utilitarian will point to the fact that the consequences of doing so will maximise well-being, a deontologist to the fact that, in doing so the agent will be acting in accordance with a moral rule (such as deal with others as you would expect to be dealt with), and a virtue ethicist to the fact that helping the person would be charitable or benevolent. There is no clear demarcation and an EHO acting to improve that person's living conditions can be said to reflect all the approaches, but that can only be said to be true if the work to address the threats to health match what the occupier themselves wanted. Actions that do not match what the occupier wanted are not benevolent nor will they maximise well-being even if the action is in accordance with the law. It can be seen therefore that the way EHOs/EHPs operate does raise ethical questions and perhaps should be reflected in guidance, recognising also that perfection can be the enemy of the good. It also militates against a "tick box" or mechanical approach that does not satisfy the needs of those affected.

As an alternative situation, what if a neighbourhood has many food outlets and licensed premises many of which are substandard so far as noise and food safety are concerned, such that it would be possible under the law to prosecute or close some down and thereby change the nature of the neighbourhood. Would closure be the ethical choice given that so that many people on low wages would lose their jobs and income, or is a more pragmatic approach, providing help and advice on securing improvements, but which would take time, more ethical? Do professional codes of ethics help answer such questions? If not, should they? If so, does that go beyond professional behaviour and being conscientious?

Professional Codes of Ethics

The first chapter ended with mention of the Nolan Principles.[20] These are:

- Selflessness (acting solely in the public interest)
- Integrity (avoid putting themselves under any obligation to people or organisations that might try inappropriately to influence them in their work)

- Objectivity (decisions taken impartially, fairly, and on merit using best evidence and without discrimination or bias)
- Accountability
- Openness (open and transparent decision-making with information not withheld from the public)
- Honesty (truthful)
- Leadership

It should be noted that although we hear a lot about these principles in the context of MPs' behaviour it applies equally to local government officers.

Many professions including the environmental health profession have members whose employment is outside any form of government. This is one reason why professional bodies have their own code of ethics. As has been made clear, a code of conduct or ethics is an essential feature of any profession and underpins the integrity of that profession. This allows self-regulation to justify a licence to practise, although EHPs/EHOs are registered and not licensed.[21]

How effective these are in regulating the behaviour of members relies both on effective enforcement including a rigorous complaints procedure, and the extent to which members of the profession must be members of the professional body or registered to secure employment. If regulation and the employment market mean that such membership or registration is not required, then arguably the whole code is undermined. While many codes govern the conduct of members they do not always make clear what the course of action should be where there is conflict between the code and what is being required of the individual member of the profession. This indicates the need for a system of advice or arbitration.

United Kingdom

The CIEH has a code that is available to the public on its website. Its code of ethics and fitness to practice rules states

> *In making decisions on what constitutes ethical practice, members will need to consider the application of technical competence and the use of their professional skill and judgement. They should also be mindful of the importance of fostering and maintaining good professional relationships with the public and other professionals.*

While this code is mandatory and sets out the principles members must follow, as we have seen not all EHPs are members, indeed many potential members choose not to be. Currently, the majority of EHPs outside Scotland will be registered under the old regime of the EHRB.[22,23] Could it be that registration does not confer sufficient advantage when it comes to employment, particularly in local government, even if under the new regime maintenance of registration requires maintenance of CIEH membership and so adherence to the code of ethics? Do prospective employers/clients even check their registration status? Another consideration is that it might be that greater professional independence might also make the employee less biddable to politicians and higher management.

This code is based on four ethical principles, which constitute the main domains of responsibility within which ethical issues are considered. These are:

1 Integrity
2 Competence
3 Responsibility
4 Respect

Some aspects of each of these headings are examined later in the chapter after reviewing the ethical requirements of environmental health bodies around the world and other professions.

There is no requirement to sign up specifically to the CIEH code, it is inherent in joining the organisation. Discussions with a representative of the CIEH[24] indicated that attention is drawn to the byelaws, the code and Fitness to Practise provisions which are on the website and publicly available when joining, but there is no requirement for members to demonstrate that these have been read. Indeed, there is no certainty that these documents are read, although they are referenced when the CIEH representatives speak to students on the environmental health degree courses.[25]

In Scotland, a conscious decision has been made not to publish a code on the Royal Environmental Health Institute of Scotland (REHIS) website and there is no specific "code of ethics" for EHOs in Scotland but there are regulations covering Professional Conduct and Discipline.[26] This requires members to act with professional competence, diligence, and honesty "and shall seek to uphold the highest standards of Environmental Health practice without regard to his own interests or to any consequences to himself or to any other person". Members should not improperly communicate privileged information; at all times uphold the law relating to environmental health information; uphold the principles of equality; and not allow professional judgement "to be influenced by political, religious or commercial consideration". The regulations address both protection of the reputation of REHIS and how members may describe themselves (post-nominal) and the disciplinary procedures for dealing with complaints (few of which have been made, and none have gone beyond the screening stage).

It should also be noted that there is also the "Ethical Standards in Public Life etc. (Scotland) Act 2000", which establishes a model code of conduct for council members. The key principles of this are: a duty to act in accordance with the law; selflessness (act in the public interest); integrity; objectivity; accountability (to the public) and stewardship; openness; honesty; leadership; and respect.

The Public Health (Scotland) Act 2008 and designation of competent person under Designation of Competent Persons Regulations 2009 make specific mention of EHOs with the designation of a local authority competent person as a qualified EHO meeting the requirements of REHIS. EHOs in Scotland also have a more obvious and greater public health role than in England. One might think that this imposes an onus on EHOs to be members of REHIS which can exercise greater control over the conduct of these professionals. However, not all members of the profession in Scotland are members even if every authority has to employ a qualified EHO. It seems that if there is an EHO as a chief officer, then a greater proportion of EHOs in that authority are also members of REHIS.[27] It is not apparent that the legal requirement gives REHIS greater power over the ethics and professional standards of officers. In contrast to England Wales, and Northern Ireland and the CIEH, control over qualification of food safety officers has been kept within the remit of a separate entity[28] administered by REHIS. The code of conduct is reinforced annually as part of the continuing professional development (CPD) regime.[29] It seems that unlike England but like the other nations of the United Kingdom, the professional body has closer ties with the devolved government and its departments, but this has more to do with the size of the countries and number of local authorities than ethics.

National Environmental Health Association, USA

The National Environmental Health Association (NEHA) in the USA has a code that requires a member to print and sign saying,

> As an environmental professional, credentialed[30] by the National Environmental Health Association I hereby acknowledge, accept, and agree to abide by the following code of conduct and ethics.

What it says is included in the boxed text below.

- I shall endeavour to keep myself current and informed and satisfy any continuing education requirements that may be in effect for my credential[31]
- I shall conduct myself in a professional manner befitting of my credentialed status
- I shall proudly represent my credentialed status to the public I serve
- I shall do nothing to undermine, detract from, or otherwise cause to develop any damaging associations with respect to this credential. I accept that any activity on my part that will cause this credential any measure of injury serves as a breach and a failure on my part

to uphold this code of ethics. Moreover, I accept that such actions for which I might be responsible could result in the revocation of my credential

- I shall do nothing to impair my ability to discharge any administrative or regulatory duty related to my professional credential that may also be required under federal, state or local law as a part of the position I hold.

South Africa

The inception of the democratic order in South Africa in 1994 has created awareness among people about the importance of human rights enshrined in the Constitution of the Republic of South Africa. Ethics also became a part of the service delivery package in the public sector.[32]

In South Africa, it is not the association itself that sets the ethical standards, but the Health Professions Council of South Africa (HPCSA) established under The Health Professions Act 1974 which is a statutory body, is the regulatory body of the profession and makes it mandatory for EHPs to uphold, promote, and maintain high standard of professional ethical conduct. The following practitioners are registered under the auspices of the Professional Board for Environmental Health Practitioners that is established under the Council, namely:

- Environmental Health Practitioners
- Student Environmental Health Practitioners
- Food Inspectors
- Environmental Health Assistant

Ethical conduct of Environmental Health Practitioners is seen as a constitutional imperative[33] and the "fundamental duty of Environmental Health Practitioners (EHPs) is to serve humankind without prejudice, to safeguard public health, and to protect them against diseases".

The HPCSA says that to achieve the standards of ethical conduct, "it is imperative for EHPs to maintain good relationships and take (sic) the interests of the clients in the execution of their duties. Ethical conduct by EHPs inculcates respect and enhances a good public image of the profession".[33] It is on this basis that the HPCSA requires all EHPs to acquire Continuous Education Units (CEUs) on ethics and on moral and ethical conduct.

It should be noted that the South African Institute of Environmental Health (SAIEH)[34] also has key principles of ethical behaviour:

- Integrity
- Inspirational standards

- Responsibility
- Respect
- Fairness
- Honesty and loyalty

Jamaica

The Jamaica Association of Public Health Inspectors (JAPHI)[35] is the organisation with the mandate to protect and enforce the interests of Jamaica's Public Health Officers, and one of its objectives is to "Police the profession to ensure that the highest level of professional and ethical standards are upheld". However, the website does not provide any further public information on a code of ethics. After training to practice, graduates must register with the Council for the Professions Supplementary to Medicine (CPSM). The CPSM is a statutory body established under The Professions Supplementary to Medicine Act 1965 and is responsible for the regulation of public health inspectors (PHIs). Its mission is to improve healthcare quality through effective regulation, CPD, and the promotion of best practices within the healthcare sector including PHIs. The Council also advises the Minister of Health on matters relating to supplementary health professions such as PHIs. It has the power to strike health professionals from the Register when they do not comply with renewal and suspend a member found guilty of professional misconduct or criminal offence. There does not appear to be any publicly available code of ethics or ethical standards for the Council or the JAPHI.

All Professionals registered under the CPSM such as PHIs must renew their practice licences annually. There are CPD provisions relating to annual renewal of the licence which also allows professionals to earn the Credit Hours needed for annual renewal. For PHIs this is 15 hours.

Canada

In Canada, the professional body is the Canadian Institute of Public Health Inspectors (CIPHI)[36] and they use the term environmental public health professionals (usually just PHIs) and EHOs. CIPHI grants the Certificate in Public Health Inspection (Canada), or CPHI(C), to candidates who fulfil the requirements established by the Board of Certification (BoC). The CIPHI code of ethics is available on the page of the website for Association Documents containing National Operational Policies (NOPs) and By-Laws, this even includes a form for complaints under the code. The code requires that all persons "applying for membership or renewing" in CIPHI must "acknowledge the CIPHI code of Ethics (Code) and agree to conduct themselves in accordance with the fundamental principles and ethics in the code".[37]

All members of CIPHI

> have moral obligations relating to their professional practice in return for the trust given to them by society. They are obliged to uphold the law and to act to protect the public's health. Members are required to follow the Bylaws and the Standards of Practice of CIPHI.

The principles are discussed in a little more detail here as they may be useful for other professional bodies and their members.

The seven principles are:

1. Justice: members should respect that the enjoyment of the highest attainable standard of health is one of the fundamental rights of every human being without distinction of race, religion, political belief, or social distinction. EHPs shall uphold the principles of justice by safeguarding human rights, equity, and fairness by promoting the public good.

   It is interesting that it says EHPs must always consider the concept of justice when performing their duties which should be performed "in a moral and righteous manner" which is perhaps quaint wording. This also involves identifying and acting to mitigate any conflicts of interest while also maintaining confidentiality of clients, colleagues, their employer, and CIPHI.
2. Inclusion: all members have equal opportunities regardless of ethnicity, race, national origin, sex, gender identity, age, ability, religion, sexual orientation, marital status, and parental status; and that they are free from harassment and discrimination. CIPHI encourages the inclusion of all people in its operations and events. inherent diversity is an important part of what makes our professional organisation so rich and robust. Their "differences mirror the diversity of our client base and bring us the skills and talents of a knowledgeable profession that is in touch with the people we work so hard to keep healthy". CIPHI encourages individuals to participate fully in all CIPHI operations and events that they wish to. Discriminatory or oppressive behaviour will not be tolerated.
3. Accountability: members "are accountable for their actions and answerable for their practice". Individuals must "practice with integrity, honesty and loyalty to CIPHI and the profession". This includes "loyalty" to CIPHI, this might be seen as a conflict with an ethical approach given that this implies that members cannot question actions of the professional body. However, an important aspect of this principle is that members should work to identify their limitations and seek help when needed, ensure that personal issues do not compromise professional performance, and should report any known violations of the code of ethics.
4. Privacy and Confidentiality: CIPHI members are expected to safeguard personal health, family, corporate, and community information obtained within the professional relationship, adhering to privacy law and preserving people's right to privacy.

5 Evidence-Informed Decision-Making: This is an interesting one – and might be assumed to be a given but it does not always work in practice in England and Wales.

> Members have an obligation to the sciences and arts for the advancement of the field of environmental public health. They shall continuously search for truths, refine their professional practice through ongoing inquiry, dialogue and reflection, and disseminate their findings for the benefit of all.

   One of the issues in the UK is how little research undertaken by the profession. The CIEH *Journal of Environmental Health Research* was relatively short lived. The CIEH did establish "a task and finish group to develop a coherent CIEH research policy". This was reported in 2012, but little was done about its recommendations.

6 Health, Well-Being, and Collaboration: every member is expected to collaborate with other "health professionals and stakeholders to achieve the best possible outcomes for clients and communities". On that basis they should act for the safety and welfare of the public and work collaboratively with other agencies and health professionals to "provide the best possible outcomes for the population of their communities".
7 Competent Practice: members, other than those whose membership status does not require it (e.g., students and affiliates), must participate in the continuing professional competencies programme. They must "strive to be current in their professional knowledge and recognise its relationship to practice" and "strive to keep themselves fully informed of the broader advancements in the field of public health".

The Canadian code is one of the most comprehensive and contains elements that others might wish to include or make their own more explicit about.

Australia and New Zealand

Australia (Environmental Health Australia – EHA) and New Zealand associations for EHOs do not appear to have any code of ethics for members, although there is a code for board members of EHA. This is available on the website. If these codes of ethics do exist, then to be of value, it can be argued they should be easily accessible to the public via the website.

Other professional bodies and ethics

In this section, we take a brief look at two bodies unrelated to environmental health to compare their approach and how their codes operate. This is intended by way of comparison.

Chartered Institute of Personnel and Development

The Chartered Institute of Personnel and Development (CIPD) provides an interesting comparator for the CIEH, although it might not be apparent at first sight.

CIPD began in 1913 as the Welfare Workers' Association (WWA) concerned with the working conditions of female employees in the UK's factories. This was a small organisation supported by employers with a strong sense of ethical obligation, Quakers being a key influence. The First World War (WW1) brought a huge increase in the number of women employed in factories in the UK. In 1916, it became compulsory for welfare officers to be appointed in all establishments controlled by the Ministry of Munitions. By the end of WW1 around 1,000 welfare workers had been appointed of whom 600 were members of the WWA. Statutory recognition gave a major impetus to growth in membership. CIPD has since evolved through a series of mergers and name changes to become the de facto leader in its field. CIPD has grown in scale continuously. By 2001 as the result of these mergers there were 110,000 members worldwide and 162,000 by 2024.[38] Change has frequently been prompted by legislative requirements, which has led to pragmatic decisions about the purpose and title of the Institute.

It has defined the emerging profession of "people management" and protected its membership's interests. Its qualifications map the skills, knowledge, and competencies required by employers and the employment environment. It has navigated its way through radical changes in industrial philosophy and structures. It survived the changes in industrial relations during the deindustrialisation of the late 20th century and is now engaging in the codification of good practice and management of remote working. CIPD has found opportunity in the changes to its environment rather than seeking to retain what were its historic strengths and boundaries.

CIPD members face similar ethical conflicts to EHOs/EHPs in that they have a responsibility to the law and to ensure that the employer understands their obligations and is compliant. They have a responsibility to act honestly in determining the truth of fact and perceptions, often reporting to outside bodies such as tribunals under oath. They are also charged with making judgements on matters of fact. They are expected to act in the context of business realities, making proportionate decisions which reflect the possible impacts which may result. As is apparent from studies on how EHOs/EHPs work in practice there is a predisposition to bring about change in behaviour rather than to take punitive action straight off. Taking housing conditions as an example, these days EHPs/EHOs typically seek to have properties restored to habitable or safe conditions standards rather than condemning the use of property by closure or prohibition or demolition and indeed with a few exceptions seemingly seek not to penalise landlords straight off as "informal action" is a common approach.[39] They are still acting within what they perceive as framework of the law and derive authority and legitimacy from it.

The standing of CIPD members has been reinforced by the extension of the law into areas such as fairness, discrimination, equality of opportunity in respect of gender, creed, and race. These days it is almost as unthinkable for a company to act without a qualified human resources specialist as it is to be without a qualified accountant. The risk to reputation and of financial retribution is too great.

The existence of a robust ethical code tailored to the professional context affords some protection against malign allegations from aggrieved persons. It has been a contributory factor in the standing of the profession which has contributed to the substantial growth in membership but there is no indication that members joined because of the existence of the code but are bound by the code of conduct and ethics. Alleged breaches of the code are dealt with by a Conduct Panel. Sanctions are available to constrain members who act in breach of the code, not least to protect the collective reputation of the profession but also to give the public confidence. The code also provides a legitimate boundary for excluding people from the profession with a history of malfeasance.

The code itself makes clear[40] members are often called upon to make difficult decisions and must think very carefully of the implication and potential consequences of decisions and actions taken. The code states that it is not exhaustive, because it is impossible to address every circumstance. Members should regulate their practice and be mindful of the code when using their own judgement and common sense, within a professional context.

The headings in the code setting out what members should do are:

1 Positive and active impact on working lives: with the purpose of championing better work and working lives for current and future generations. By promoting people-centred practice, developing people, being the employee experience champion, supporting the health, education, and welfare of those in the work environment.
2 Civic virtue and stewardship: by being socially responsible, supporting future organisational needs, and helping to improve the communities in which members live and work.
3 Good character: by applying sound judgement; being honest, not allowing personal interest to undermine the member's objectivity and protecting the reputation of the profession.
4 Professional service and competence: by providing high standards of competence and service and committing to the continuing development of the member's skills, knowledge, and experience.
5 Personal responsibility: by taking responsibility for actions, considering their impact on others, and showing courage in their approach to difficult decisions and ethical dilemmas.

Interestingly, the document has been reviewed by the Institute of Business Ethics and was benchmarked against other professional membership associations, the CIPD worked with the Professional Associations Research Network (PARN) the University of Birmingham's Jubilee Centre for Character and Virtues, and the University of Leeds's Inter-Disciplinary Ethics Applied (IDEA) Centre.

The parallel development of a qualification structure and detailed specification of the knowledge and competence required of a member of the Institute form the basis of the licence to practice. CIPD has formed alliances with other institutions such as universities to validate the equivalence of qualification and provide a ladder of progression. The qualifications for entry to the profession require broad-based tested knowledge and expertise forming the basis for future career development and progression.

The growth of CIPD membership is related to the perception that membership gives a substantial degree of licence to practice. As employment legislation has strengthened and extended over time, employers sought to demonstrate that they were appropriately briefed and compliant by hiring qualified staff who are bound by the ethical code of behaviour. CIPD members are required to demonstrate that they have the necessary knowledge and professional competence, acquired by initial training and subsequent CPD, to be able to give authoritative advice on legal requirements and industry best practice.

CIPD has invested considerable resources in building prestige and reputation and is seen as the go-to point for authoritative commentary on good practice in the workplace. This in turn adds substantial value to the individual member.

Building control

Building Control is also an interesting area as this has undergone major changes since the Grenfell Tower fire and the regulation of the Building Control profession is now under the Building Act 1984 as amended by the Building Safety Act 2022. The Building Safety Regulator (BSR) oversees building control bodies (local authorities and registered building control approvers) and registered building inspectors (RBIs). The Health and Safety Executive is appointed as the Building Safety Regulator (the regulatory authority) for RBIs by section 58A of The Building Act 1984 (as amended). The code of conduct is prepared and published in accordance with section 58F of the Act which requires the BSR must publish a code that sets out the standards of professional conduct and practice expected of RBIs. As part of the application to register the BSR will ask the applicant to confirm that they comply with the code of conduct for RBIs in the countries in which they are registered.

RBI is a protected title. It is a criminal offence to impersonate, or to do anything which would imply a person is an RBI (with the intent to deceive). The code is central to the regulated building control profession. One in which inspectors and where applicable, self-employed contractors are required to take individual responsibility and accountability for decisions, actions, and behaviour. In line with this, the building inspector competence framework (BICoF)[41] has been developed to set out the standards of competence as an RBI. Interestingly, an RBI does not have to be a member of a professional body but should include any such membership on an application to register. The regulatory authority has the power to set conditions on an individual's registration under the 1984 Act (as amended). Conditions may be imposed at registration, when a registration is varied and/or as a result of sanctions and enforcement action by the BSR.

The code[42] contains both principles and standards of conduct and are publicly available on the government website. The principles in the code are:

1 Act with honesty
2 Act with integrity
3 Maintain your professional competence
4 Deliver services with professional skill and care
5 Uphold public trust and confidence in the provision of services and the profession
6 Treat everyone fairly and act in compliance with your legal obligations

The standards to be met include a number of the more detailed requirements. First, the need to comply with legal, regulatory, and professional obligations including the need to comply with industry standards and consider good practice where appropriate. The RBI must maintain professional independence and impartiality in the conduct of work activities. It also requires RBIs to take appropriate steps to ensure that work undertaken by individuals under their supervision complies with relevant legal and regulatory requirements.

Workplace requirements include an interesting provision on whistleblowing so that RBIs must familiarise themselves with their organisation's whistleblowing policy and report concerns in accordance with it and must not act in a way which prevents or discourages whistleblowing. They are also required to consider whistleblowing concerns that are received fairly and keep records for at least 15 years from the date of report. As with many codes there is a requirement to act in a manner that supports a culture of inclusion, equality, and diversity in accordance with their organisation's policy.

On maintaining professional competence and CPD, RBIs must only carry out restricted activities for which they are registered and have the necessary competence, and the same for managers or supervisor. They must take steps to ensure the ongoing competence of those they manage or are under their supervision. Of relevance to EHPs/EHOs is the requirement to "formally reflect on your development needs and record your findings and planned actions to meet these needs in accordance with the continuing professional development requirements". They have to maintain a portfolio of evidence to demonstrate their competence including records of training, development, and competence activities that have been undertaken.

When it comes to the standard of service RBIs are required to conduct their work, fairly and objectively, diligently, conscientiously, and in the best interests of the public when dealing with individuals, other professions, or the building control profession. The code has a number of Annexes. Annex 1 covers obligations to the regulatory authority under this code, and Annex 3 relates to registration conditions for RBIs.

As we have seen there is no requirement for RBIs to be members of a professional body, but it is useful to look briefly at the Chartered Institute of Building (CIOB).[43] This has grown out of the first group of Master Builders which came together in 1834. The CIOB was granted a Royal Charter in 1980 and has over 49,700 members in more than 100 countries.

The website states that "Ethics and ethical behaviour are central to professionalism. It does not matter how skilled and experienced a person becomes: if they behave dishonestly and without regard for the rights of others, they are not a professional person". Its objectives, as defined in its Royal Charter, are the promotion for the public benefit of the science and practice of building and construction and the advancement of public education in the said science and practice including all necessary research and the publication of the results of all such research.

The website reports the 113th President of the CIOB's inaugural address to politicians and industry leaders in 2016,[44] where he said one of his objectives was to make ethics a central theme for the year ahead saying "Ethics defines what it means to be a professional. Behaving ethically entails tough decisions. And we have a responsibility to help our members make those right decisions".

Chartered Trading Standards Institute

Founded in 1881, the Chartered Trading Standards Institute (CTSI) is one of the world's oldest organisations focussed on Trading Standards and Consumer Protection. The code is one page so has the benefit of brevity.

Under the heading of "professional competence and behaviour" members owe their professional duty in all individual professional matters to their employer (or client if self-employed) and should fulfil that duty to the highest professional standards. Where a conflict arises between the requirements of an employer or client and the aims and objectives of Institute, then the CTSI should be notified. The CTSI shall be notified of the outcome of any legal proceedings that may bring their character or professional standing into question. Members are also required to maintain professional knowledge and competence through CPD to ensure they can provide an "up to date and competent service".

On ethical standards and integrity members shall always be honest and courteous, champion professional practices that promote equality of opportunity, diversity and inclusion and support human rights and dignity. They shall also demonstrate and promote sensitivity for the customs, practices, culture, and personal beliefs of others. Members should not accept any preferential loans, gratuities, "or other benefits, which could compromise or potentially compromise their professional integrity, and which could be held to have been offered solely as a result of their professional position".

Members as representatives of the profession are required to act in a way which supports and upholds the reputation and values of CTSI and the Trading Standards profession, report any potential breach of this code, uphold their responsibilities as professionals towards the wider community, comply with prevailing laws and not encourage, assist, or collude with others who may be engaged in unlawful conduct. They are expected to be a role model for maintaining the highest standards of ethical conduct.

On professional standards, members shall treat all people, colleagues, clients, and any other stakeholder, with respect, acting fairly and reasonably at all times. They have to be prepared to challenge others if they suspect unlawful, unethical or inappropriate conduct or behaviour, taking action or raising concerns as appropriate. Their professional judgement must not be compromised nor perceived as being compromised because of bias, or the undue influence of others. They should also assist fellow Members and wherever possible share information and expertise which may be of assistance in the proper performance of their statutory duties, or the objectives and interests of the Institute.

Institution of Fire Engineers

The Institution of Fire Engineers (IFE) is the professional body dedicated to "promoting excellence in fire engineering" and was founded in 1918. It is a registered charity and has about 11,000 members worldwide. It has a code of conduct for its members and in addition to complying with this code members are expected to comply with four ethical principles.

Under the code of professional conduct[45] IFE members of all grades shall

- Uphold the "reputation and standing of the Institution and the profession of fire engineering"
- Observe the provisions of the Institution's governing documents and supporting rules and regulations
- Accept responsibility for work carried out by them, or under their supervision exercising all reasonable professional skill and care in carrying out that work
- Hold adequate professional indemnity insurance, where required

- Exercise all reasonable professional skill and care to prevent avoidable danger to health or safety, and to prevent avoidable adverse environmental impacts
- Maintain their professional competence and only undertake professional tasks for which they are competent
- Avoid real or perceived conflicts of interest and advise affected parties accordingly
- Discharge their professional duties with honesty, integrity, impartiality, and objectivity
- Observe the duty of confidentiality owed to appropriate parties
- Support any colleague, or any other person to whom they have a duty of care, who in good faith raises any concern about a danger, risk, malpractice, or wrongdoing which may affect others. The IFE has a Whistleblowing Policy
- Notify the Institution of any serious or significant violation of this code of professional conduct by another member. Reference is again made to Whistleblowing Policy and
- Notify the Institution immediately of any criminal conviction, or an adverse civil court judgement, or been declared bankrupt or disqualified as a company director or charity trustee or had membership of another professional body or institution terminated as a result of a disciplinary procedure.

The whistleblowing policy says that while whistleblowing is not a legal requirement, the IFE believes that members should understand their professional obligations with respect to whistleblowing[46] and know where to go to for support when they have a concern that they believe should be raised. This is available on the website,[47] and this is not covered in any other codes seen while preparing this book and is perhaps an omission that should be covered in guidance on professional ethics. It seems to be an important aspect of ethical behaviour to be able to alert others to ethical problems and issues.

The ethical principles for the IFE as set out are:

1. Honest and integrity
2. Respect for life, law, the environment, and public good (this includes drawing attention to any hazards that pose a threat to health and safety)
3. Accuracy and vigour (members should keep their skills and knowledge up to date and assist in the development of knowledge and skill in other professionals)
4. Leadership and communication

Included above in brackets is some of the supporting text that is of relevance to EHOs/EHPs.

The IFE is of interest because it came into the news at the time of writing this book and it made the housing press[48] when a member who had undertaken work for housing associations was expelled from membership for non-compliance with a sanction. The company was suspended from the IFE in August 2024 for breaching the code of conduct and also the organisation's ethical principle of "accuracy and vigour".[49] The expulsion was the result of failing to comply with the original sanction of a six-month suspension from membership and registration. Tri Fire had carried out risk assessments and External Wall System 1 (EWS1) surveys and fire engineering on dozens of buildings. Another member was expelled in February 2025 for non-compliance with the sanction. One member who has been sanctioned and is on the list of sanctioned members chaired the IFE's board of trustee directors until summer 2023.

At a time when fire safety concerns are perhaps greater than they have ever been it is important that housing providers, owners, and local authorities (including EHOs/EHPs involved in housing enforcement) need to have confidence that those firms and individuals who provide "expert" reports are reliable. That is what membership of a professional body should provide.

CIEH code and fitness to practise in more detail

Having looked at other professional bodies we return to consider the code of the CIEH. The current code is publicly available on the website and is that Agreed by the Board of Trustees on 30 November 2023 and to be reviewed before 31 December 2025. It is important that, as a general rule, such documents are regularly reviewed and revised accordingly, and that the most up-to-date version is freely available.[50] The code has generally accepted principles set out above but has some detailed "behaviours" that indicate compliance with these.

It makes clear that applying the first principle (integrity and protection of the public is the prime concern) members must put the interests of the communities they serve first before their own interests and those of any colleague or organisation. Information and advice that members provide to the public, employers and colleagues shall be prompt, clear, and accurate. Members should take swift and appropriate action if aware that their health, behaviour, or professional performance, or those of a colleague, or the actions of an organisation, may compromise the health of the public. Where a member is unsure how to act in a particular situation, they shall seek advice from "appropriate colleagues" or a professional organisation. What are "appropriate" colleagues is unclear – what if such advice is contrary to the principles of the code? Being a CIEH code, surely that should be the named professional organisation that provides advice or if there are alternatives would it not be better to specify criteria if not specific organisations?

Is this also something that could be covered in a policy on "whistleblowing"? Importantly if a

> *member perceives a conflict between two or more principles in this code, or between them and any other applicable guidance, they shall take the course of action that they judge most likely to protect the public and promote environmental health.*

What it does not say is that any perceived conflict should be reported to the CIEH. One might have thought that knowledge of such conflicts will help the professional body improve the code and provide advice. The professional body should be available to provide advice on ethical matters.

Part of the first principle is being honest and trustworthy "straightforward, honest and fair" unbiased in their professional judgements and "not be associated with reports, returns, communications or other information where they believe that the information contains a false or misleading statement" or "omits or obscures information required to be included where such omission or obscurity would be misleading" and not knowingly mislead anyone. In the same tone and recognising that some EHPs are engaged in research (probably not as many as should be) they should apply best evidence honestly and impartially. That opens up a number of questions and with EHPs working as consultants, in contract as well as academic research and "experts" as well as in local authorities it is surprising there are not more complaints. All offers of gifts and hospitality that may affect, or be perceived as affecting, their judgement should be refused. Conduct that could affect or undermine the confidence placed in them, the CIEH and the environmental health profession should be avoided. The second part of the first principle is fundamentally about being honest (including in financial matters) and trustworthy and the first principle picks up two of the Nolan Principles.

The second principle (competence) is all about maintaining high standards of professional and personal conduct and knowing one's limitations. This includes "proper standards of work and keeping accurate records". "Proper" is not defined and raises a question of how that can be gauged where, for example, differences of "professional judgement" arise. Conflicts of interest that may arise between professional work and the health of the public that should be avoided may be easier to identify. Again, could there not be a conflict of interest when it comes to charging for re-inspections for "scores on the doors" food hygiene ratings where a local authority as the result of cuts is trying to maximise income? In the same way, the fees for licensing under the Housing Act 2004 and temptation to maximise fee level to increase income when funding for the environmental health service is inadequate.[51] In some ways, licensing fees may seem a more attractive administrative proposition than use of enforcement powers and the civil penalty regime for non-compliance. There is also then the issue of how such civil penalties are used and whether they are used correctly by the local authority.[52] What should the EHO/EHP do if it appears that these are not being used as intended by the legislation?

Of great importance and possibly one that is breached more often than admitted is the need to develop and update professional knowledge and skills throughout one's working life and undertaking relevant training and learning about best practice. This must be a prerequisite of "competence". It is also required that members should keep their "knowledge, skills and professional performance under continuous review, reflecting on them systematically to identify strengths and weaknesses and complying with all requirements for continuing professional development". For this provision to be applied effectively, greater quality control of CPD hours is needed than appears to be the case. It is known from a study of an area of environmental health work that some officers have indicated they have no need of further training.

It is a requirement that members should decline to act if they lack the knowledge, skills, experience, or authority to undertake a piece of work. They should refrain from providing advice where they have reason to believe they lack the professional knowledge and skill at the level required. This is an interesting one as if a member has failed to keep their knowledge and skills up to date, then could they be said to be in breach of this requirement as well – where is the boundary? For example, the statutory Operating Guidance for the Housing Health and Safety Rating System also requires users to keep up to date with housing and health research – if they fail to do so yet still carry out assessments that lead to action under the Housing Act 2004, are they in breach of the code?

The code further explains this principle by saying members should be prepared to refer a matter to an appropriately qualified and experienced colleague. Nor should they present themselves as having a qualification, grade of membership, designation, or experience that they do not. In what is something of a matter of safeguarding the organisation, the fitness to practise provisions say members should not put forward views on behalf of the CIEH without the authority to do so.

The third principle (responsibility) is rather about protecting confidentiality, so that information learned about individuals in the course of their work should remain confidential with information disclosed only to those entitled to receive it – one assumes legally entitled. Members should ensure safety of documents (paper or electronic) which should be stored and transmitted securely. Confidential information acquired as a result of professional relationships should not be used for the personal advantage of the member or anyone else. One could argue that this is also a matter of honesty and integrity.

Finally, the fourth principle (respect) covers respecting the dignity of individuals and treating everyone fairly and co-operating with others. It might seem obvious, but members should treat everyone politely and with respect (that word again) recognising the individual's right to make choices and be involved in decisions which affect them. One must wonder to what extent tenants are involved in decisions which affect them when local authority EHPs inspect their homes, identify hazards or other problems and then take action under the Housing Acts to secure a remedy, particularly when there are options on the course of action to be taken.

Everyone should be treated equally regardless of their gender, race, nationality or ethnicity, disability, age, sexual orientation, transgendered status, religion or belief, marital or civil partnership status, recognising differences between individuals and groups, and avoiding stereotyping. Members are expected to listen "to individuals, groups and communities, and give them the information they need in a way they can use it" and not abuse professional relationships tolerating other members of the CIEH, respecting their views and perspectives on professional and technical matters.

This principle extends beyond other CIEH members and into the workplace by expecting cooperation with other "team members and colleagues". At the same time communicating effectively and sharing knowledge, skills, and experience with colleagues, employers, and others in the interests of the public and assisting colleagues in complying with requirements for CPD. Members should also "provide proper supervision of tasks they have delegated to others, recognising that they remain accountable for work they have delegated". In the workplace, members should behave with "honesty and be impartial in assessing someone's suitability for employment the performance of someone they have trained or supervise". It is unclear how Members can always be expected to make sure there is an effective complaints procedure where they work if not in a managerial position, and then follow it at all times, cooperating fully with those investigating or adjudicating upon a complaint.

This set of "behaviours which we treat as the standards of conduct, setting out the professional conduct that the CIEH expects of its members" is very detailed, and it is somewhat surprising that there are relatively few complaints made in any one year – even if they do not lead to any sanction. There is a question as to whether such a detailed set of behaviours achieves its aim. Would a simpler and shorter code be any less effective? Does this code make the profession any more ethical and behave better than other professions? Is the extensive list of behaviours an example of mission creep? Does it address adequately how ethical dilemmas faced by EHPs should be addressed and what support it should give to members. It is all in one direction, saying how members should behave but does not address what the Institute can do help members.

In what seems to be a unique approach the publicly available code also sets out the procedure for complaint handling and any member of the public may bring to the attention of the Chief Executive any facts or matters indicating that a member may have breached the code and become liable to fitness to practise proceedings. Members themselves "are under a general obligation" to notify the Chief Executive matters that might indicate a breach and thus indicating that another member may "have become liable to fitness to practise proceedings". Members who have been convicted of any criminal offence "in any jurisdiction" should also notify the CIEH.

The detailed procedure for dealing with complaints includes initial screening by a "Screener"[53] who can investigate the complaint. If the Screener determines that there is a prima facie case to answer, the case shall be referred to a Fitness to Practise Panel. If the Screener determines that there is no case to answer the matter shall be referred to a Lay Moderator. The Lay Moderator shall consider the decision made by the screener and shall confirm or overrule that decision. In the latter instance, the case is referred to the Fitness to Practise Panel and a hearing convened.

There are rules governing how the Fitness to Practise Panel operates which are available on the CIEH website. There are limited grounds for appeal to an Appeal Panel. The grounds are unreasonableness of the decision, or there is additional evidence not available at the time of the hearing, or there was a material deviation from the regulations governing how a complaint was dealt with, or the proposed sanction is disproportionate.

As has been said there are not many complaints under the code to the CIEH in any year – less than 20 complaints and less than ten reach the panel. Most come from the public about a member (or their local council) rather than from a member (or the CIEH itself) about a member.[54] A large proportion of complaints may be dealt with at the screening stage as just three hearings in 2023 and 2024 (plus on appeal relating to one of the hearings) are reported on the website. That said only hearings where a membership is terminated are published, unless in other cases the respondent specifically asks for publication of the hearing. The hearings are in public, and the decisions of the panel are published on the CIEH website.[55] In one such decision, the CIEH member had claimed to be a Chartered EHP when he was an Associate (non-voting member) and in another case a member working for a consultancy that had been contracted to carry out food hygiene inspections, had fraudulently claimed to have carried out inspections and claimed payment for inspections that had not been undertaken. In the third case, a Chartered Member was guilty of misconduct in that when undergoing CIEH's pathway to become a Registered Food Safety Practitioner the signature and name of the person purporting to verify the competencies demonstrated in their competency development portfolio were applied fraudulently without their knowledge. If there are few complaints, is this a reflection of the probity of EHOs/EHPs or that there is little recognition of what is required of the ethical professional or that indeed the available sanction makes the process inconsequential?

It is not apparent if the CIEH has had to deal with any cases where members have put the interests of the communities they serve and public health first before those of any colleague or organisation, which one assumes includes their employer. While members "shall take swift and appropriate action if they become aware that …… the actions of an organisation, may compromise the health of the public". It is not clear what that "swift action" should be and how well the CIEH will handle this? If a member is seeking to behave ethically but this brings them into conflict with their employer or the person to whom they are contracted, it is incumbent on their professional body to set out how this should be dealt with.

What is a relevant consideration is whether a member or prospective member has to acknowledge the code and standards of behaviour. Such people are sent links to the Byelaws on the CIEH website but there is no requirement for them to show they are aware of either the Byelaws or the code of ethics. When the Executive Director of Professional Standards makes presentations to members or environmental health students (potential members) the code is highlighted. In referring to the website, an analogy was drawn with on-line terms and conditions of websites or programmes whereby people scroll down and click "agreement" without actually reading them. Whether this is a reasonable or adequate approach is perhaps debateable.

There is no statutory definition of "fitness to practise". However, CIEH views fitness to practise as a member's suitability to remain as a member of CIEH without restriction and so removal from membership is the greatest sanction that can be imposed but in one of the cases above the member was barred from sitting the exam for 12 months. Removal from membership does not prevent the person from continuing to work as an EHP, whereas in other professions they would be "struck off" and would not be able to practise, for example, striking off the roll means that a lawyer is no longer allowed to practice law or a "struck-off" doctor in the UK means they have been removed from the GMC register.[56] For EHPs it is said that employers generally require Environmental Health Registration Board, CIEH, or REHIS registration for EHO/EHP posts. Registration as an EHP is now tied to membership of CIEH. The CIEH is "encouraging the terms Registered and Chartered EHP with the use of the letters REnvH and CEnvH as desirable for employers who wish to have staff tied into an ethical code and commitment to audited CPD". Some employers do have membership or Chartered Status as either essential or desirable in their job specifications.[57] There is some anecdotal evidence that some people who have had their memberships terminated have ceased to work for their employer. Reference has been made to local authorities using partly qualified or non-qualified staff to carry out environmental health functions, but it has been suggested that even if local authorities had the necessary finances there is a shortage of people qualifying as EHPs/EHOs.

There may be some scepticism as to whether a statutory status could be achieved in the current climate, and while statutory regulation (and striking off) could better ensure improved ethical behaviour it might not increase recruitment into the profession. However, we would argue that not only would it help ensure that members operated ethically but may even help recruitment onto environmental health courses and then into the profession. Yet there is a clear impression that more could be done, for example, on reserved activities and protected titles. It is beyond this book to consider routes into the profession, but having a sound ethical basis and regulatory regime is perhaps one selling point.

European Convention on Human Rights and the Human Rights Act

Codes of ethics and their requirements frequently reflect what have been taken as basic human rights whether explicitly or implicitly. EHOs/EHPs are part of the public health workforce, and public health work requires both ethics applicable to the individual public health (or environmental health) practitioner and a human rights framework to guide public health in its societal analysis and response.[58]

EHOs working for local authorities and other agencies of government are also part of the state for the purposes of the EHCR and thus should be aware of the provisions of the Convention. The Human Rights Act 1998 incorporates the ECHR into UK law. Under the Act it unlawful for any public authority (any person certain of whose functions are functions of a public nature such as a local council) to act in a way which is incompatible with the convention. It also requires the judiciary (including tribunals) to take account of any decisions, judgement, or opinion of the European Court of Human Rights, and to interpret legislation in a way which is compatible with the Convention.

This has become an issue as the result of some right-wing politicians in England wanting to abandon it and avoid being within the jurisdiction of the European Court of Human Rights (this is not the same as the European Court of Justice with which some politicians get confused and which deals with EU law). The Convention follows on from the Universal Declaration of Human Rights proclaimed by the General Assembly of the United Nations on 10 December 1948. Article 1 imposes an obligation to respect human rights. Other relevant Articles are:

- Right to a fair trial: where there is a criminal charge, everyone is entitled to a fair and public hearing within a reasonable time by an independent and impartial tribunal established by law.
- Right to respect for private and family life: everyone has the right to respect for their private and family life, home, and correspondence.
- Freedom of thought, conscience, and religion: everyone has the right to freedom of thought, conscience, and religion.
- Freedom of expression: everyone has the right to freedom of expression, including freedom to hold opinions and to receive and impart information and ideas without interference by public authority.
- Prohibition of discrimination: enjoyment of the rights and freedoms in the Convention shall be secured without discrimination on any ground such

as sex, race, colour, language, religion, political or other opinion, national or social origin, association with a national minority, property, birth, or other status.

Conclusion

What are called "codes of ethics" more often set out the standards of behaviour or conduct of members of that profession. Some elements often seem to be about safeguarding the reputation of the professional body, whereas the notion of ethics relates to society as a whole. The issue of whistleblowing is contained within a minority of codes, and this must be an important part of professional ethics along with support from the professional body. It is also unclear to what extent ethics as a whole is promoted within the different professions, or is only lip service paid to an ethical approach? Having a strong ethical code does not prevent members behaving unethically so the code has to be supported by a robust system of enforcement that is transparent.

Notes and References

1 Battersby SA, Ed. (2022) *Clay's Handbook of Environmental Health,* 22nd Edition, Routledge, Abingdon, Oxon. pp 78–80.

2 See https://www.regulated-professions.service.gov.uk/professions/environmental-health-practitioner and https://www.regulated-professions.service.gov.uk/professions/environmental-health-officer

3 Coggon J and Viems AM, (2017) *Public Health Ethics in Practice – A Background Paper on Public Health Ethics for the UK Public Health Skills and Knowledge Framework*, Crown Copyright.

4 What is meant by "environmental health" is considered at length in Battersby SA, Ed. (2022) *Clay's Handbook of Environmental Health,* 22nd Edition, Routledge, Abingdon, Oxon.

5 Riggs DE, (2016) The Environmental Health Profession. *Journal of Environmental Health*, Vol. 79, No. 5 (December), pp. 6–7, National Environmental Health Association, Denver, Colorado. https://neha.org/sites/default/files/jeh/JEH12.16-Pres-The-Environmental-Health-Professional.pdf

6 By comparison in the medical profession the role of the GMC is to 'protect, promote and maintain the health and safety of the public by ensuring proper standards in the practice of medicine'. The GMC does not itself set clinical standards – this is more properly the remit of the medical royal colleges. The BMA is in effect the trade union for the medical profession.

7 https://www.who.int/about/governance/constitution

8 Schrammer T, (2023) Health as Complete Wellbeing: The WHO Definition and Beyond. *Public Health Ethics*, Vol. 16, No. 3 (November), pp. 210–218. https://doi.org/10.1093/phe/phad017

9 *R.* v *Bristol City Council Ex Parte Sandra Everett* [1998] EWCA Civ J0513-2 - the local authority had served an abatement notice to secure replacement of a dangerous staircase, but then realised and error so withdrew the notice. This was judicial review of that decision.

10 See, for example, https://www.socratic-method.com/quote-meanings-interpretations/albert-einstein-the-environment-is-everything-that-isnt-me

11 Chanlett ET, (1958) What Makes a Profession. *The Sanitarian*, Vol. 20, No. 6, pp. 286–290, National Environmental Health Association (NEHA). https://www.jstor.org/stable/26325506

12 The difference is explained later but relates to the regulated professions – EHOs in Scotland (and other countries) and EHPs elsewhere in the UK.

13 See Szentkiraly L, (2023) Moral Foundations of the Precautionary Principle: Juul and the Hazards of Vaping, in Battersby S, Ed, *Clay's Handbook of Environmental Health*, pp. 57–63. Routledge, Abingdon Oxon.

14 See, for example, https://www.localgovernmentlawyer.co.uk/sharpeedge/809-sharpeedge-procurement/44146-local-government-contracts-and-the-ultra-vires-doctrine and https://www.local.gov.uk/introduction-localism-act and Localisms Act 2011 which introduced the general power of competence which replaced the previous power of well-being and gave councils the same broad powers as an individual to do anything unless it is prohibited by statute.

15 Hugman R, (2005) *New Approaches in Ethics for the Caring Professions*, Palgrave Macmillan, Basingstoke.

16 Lipsky M, (1969) Toward a Theory of Street-Level Bureaucracy (IRP Discussion Papers No. 48–69) (p. 45). *Institute for Research on Poverty* (IRP), University of Wisconsin. Retrieved from http://www.irp.wisc.edu/publications/dps/pdfs/dp4869.pdf

17 https://www.oed.com/dictionary/vocation_n?tab=meaning_and_use

18 https://www.gmc-uk.org/professional-standards/good-medical-practice-2024#Knowledge-skills-and-development

19 https://plato.stanford.edu/entries/ethics-virtue/

20 https://www.gov.uk/government/publications/the-7-principles-of-public-life/the-7-principles-of-public-life--2

21 Registration generally involves recording information on a registry or database to recognise existence or status or an individual or organisation, whereas a licence specifically authorises someone to perform a particular activity.

22 Prior to September 2021, Environmental Health Practitioners (EHPs) and related professionals who qualified in England, Wales and Northern Ireland were Registered with the Environmental Health Registration Board (EHRB). CIEH now owns its Registers, which are published at https://www.cieh.org/professional-development/professional-registration/pathways-to-registration/ehrb/

23 Registrations with the EHRB remain valid indefinitely. Those who also wish to appear on CIEH's Register and have access to the relevant digital credentials (and REnvH designatory letters in the case of Environmental Health Practitioners) can do so if they are a member of the CIEH have adequate CPD records and pay a fee.

24 Private discussion on 28 January 2025.

25 The number of courses and students to fill them to keep them viable is an issue for the future of the profession.

26 A copy of which (Version dated September 2024) was provided to the authors by the Chief Executive of REHIS.

27 Conversation with Jackie McCabe Chief Executive of REHIS.

28 Scottish Food Safety Officers' Registration Board (SFSORB).

29 The REHIS Scheme of Continuing Professional Development (CPD) requires participants to undertake a minimum of 60 hours of CPD over three consecutive calendar years, with a minimum of 15 hours activity to be undertaken each year. Within the CPD hours is a requirement, to submit a written submission at least once in every consecutive three-year period.

30 An American term meaning "having qualifications or documentation indicating one's suitability for something" so that inspections would have to be done by a group of credentialed inspectors.

31 Defined as the ability and experience that make someone suitable for a particular job and/or proof of their abilities and experience. (see Cambridge Dictionary https://dictionary.cambridge.org/dictionary/english/credential).

32 Poswa TT, (2017) *Environmental Health Practice in a South African Metropolitan Municipality: Professional, Ethical and Legal Responsibilities and Challenges*, A thesis submitted to the Faculty of Health Sciences, University of the Witwatersrand, in fulfilment of the requirements for the degree of Doctor of Philosophy.

33 https://www.hpcsa-blogs.co.za/ethical-conduct-of-environmental-health-practitioners-a-constitutional-imperative/

34 https://saieh.co.za/

35 https://japhi.org.jm/

36 https://ciphi.ca/

37 Canadian Institute of Public Health Inspectors (CIPHI) Code of Ethics revised June 2020.

38 Personal communication with Rose Marney Corporate Custodian and Librarian of CIPD.

39 Battersby S, (2017) *Local Housing Authority Action on Conditions in Rented Housing*. Report following a Freedom of Information request to local housing authorities prepared for Karen Buck MP. Available at https://www.sabattersby.co.uk/file-share

40 CIPD Code of Conduct and Ethics of January 2023, at https://www.cipd.org/uk/membership/professional-standards/code-of-conduct/

41 https://www.gov.uk/government/publications/registered-building-inspectors-professional-codes-and-standards/building-inspector-competence-framework-bicof

42 See https://www.gov.uk/government/publications/registered-building-inspectors-professional-codes-and-standards/code-of-conduct-for-registered-building-inspectors

43 https://www.ciob.org

44 https://www.ciob.org/news/ethics-at-the-heart-of-what-it-means-to-be-a-professional-says-new-ciob-president-paul-nash

45 https://www.ife.org.uk/Portals/0/Documents/Policies%20and%20procedures/IFE%20Code%20of%20Conduct.pdf?ver=2024–09–25–103913-050

46 A whistleblower is a worker and who reports certain types of wrongdoing. This will usually be something seen in their employment – though not always. The wrongdoing disclosed must be in the public interest. This means it must affect others, for example, the general public.

47 https://www.ife.org.uk/Portals/0/Membership/IFE_whistleblowing_policy.pdf?ver=2024–08

48 https://www.insidehousing.co.uk/news/news/fire-engineer-fully-expelled-from-professional-body-90525

49 https://www.ife.org.uk/Portals/0/Documents/Sanctions/IFE%20Summary%20of%20case%20-%20Adam%20Kiziak%20updated%2007.02.25.pdf?ver=2025-02-07–145531-537

50 https://www.cieh.org/media/3rjajx10/code-of-ethics-for-members-and-ftp.pdf

51 In *R (Gaskin) v Richmond upon Thames LB* [2018] EWHC 1996 (Admin) Divisional Court the Administrative Court has held that a person who owns, and lets out rooms in, an HMO provides a service for the purposes of the Provision of Services Regulations 2009, SI 2009/2999 and that the HMO licensing scheme is an "authorisation scheme" for the purposes of the Regulations. The decision of the CJEU in *R (Hemming t/a Simply Pleasure) v Westminster CC* [2017] 3 WLR 317, therefore applies to the fees that may be charged on a Part 2 licensing application. It is unlawful to demand payment of an upfront fee with an application which covers not just the costs of processing the application, but also the costs of administering and enforcing the scheme.

52 Guidance from MHCLG of 2018 says "Income received from a civil penalty can be retained by the local housing authority provided that it is used to further the local housing authority's statutory functions in relation to their enforcement activities covering the private rented sector, as specified in Regulations", i.e., SI 2017 No. 367 The rent Repayment Orders and Financial Penalties (Amounts Recovered) (England) Regulations 2017.

53 The "Screener" a person appointed by the CIEH for the purpose of regulation (who may be a member of CIEH staff, a CIEH member, or neither of these). Such a person may not serve as a Lay Moderator or as a member of a Fitness to Practise Panel.

54 Conversation with Jon Buttolph of CIEH.

55 https://www.cieh.org/governance/how-cieh-is-governed/governance-documents/

56 The GMC maintains a register of doctors, and a doctor wishing to practise medicine in the UK will also need to hold a licence to practise along with the appropriate registration for the work undertaken. Status on the medical register shows if that person holds a licence or not.

57 Personal communication from Jon Buttolph of CIEH.

58 Mann JM, (1997) Medicine and Public Health, Ethics and Human Rights. *The Hastings Center Report*, Vol. 27, No. 3, (May–June), pp. 6–13. The Hastings Center. https://www.jstor.org/stable/3528660

3 The law and ethics

Introduction

Although ethics and law are closely related, they are not the same and just because something is legal (or complies with the law) it does not make it ethical. It might be thought that differentiating between the two is obvious, but it is not always so clear cut. There can be actions that are not prohibited by law but that are not considered to be ethical. For example, purchasing goods that are supplied by a business that underpays its employees or causes harm to the environment or lying about something that is not illegal. Ethical considerations may exceed the need or the requirements of the law, especially in the health sector, when the law may set only minimum standards or requirements. The question is, should it be sufficient for an environmental health practitioner (EHP) to say the law "requires this" or "does not require this" if their professional judgement is different? Or to say I am implementing the law or the policy of my employer even if the circumstances might require that greater or different action or intervention be undertaken. Or is it right to say the law permits this, so this is what we will do, without any consideration of the impact on public health, health equity, or whether it is ethical. For example, looking at other professions, accountants and lawyers can give professional advice on tax avoidance but not tax evasion which is illegal and involves concealing income or misrepresenting financial information to avoid paying taxes, while tax avoidance is legal and refers to using strategies like deductions and credits to minimise tax liability. The key difference lies in legality but perhaps for some there is less difference ethically. For EHPs arguably they should also be looking analytically at the public health impacts of their actions or inactions and the effects on health inequality. For example, what are the ethical considerations associated with charging regimes[1] and the use of civil penalties particularly at a time when local government is under severe financial constraints, if such costs lead to higher rents or eviction?

Hutter[2] has highlighted the importance of the context in which environmental health officers (EHOs) operate and has demonstrated their adoption of a particular version of the accommodative approach that she has described.[3]

DOI: 10.1201/9781003625339-3

EHOs/EHPs operate in a very different environment from other agencies charged with addressing compliance (or non-compliance) with the law. There are the potential (implicit) and actual political influences of the EHOs' employers, usually the local authority. Hutter showed that the extent to which the local authority seeks to exercise control is crucial, yet the law itself may not be fully understood. For EHOs there are other social actors that have an influence on how the law is used, notably the public. These may be apathetic so far as environmental health law enforcement is concerned if not directly affected and therefore do not accord a strong mandate to the officers to adopt a stringent approach to the law. Hutter showed that there is a general ambivalence to the use of the criminal law, whereby it would seem that morality, or moral perceptions (and perhaps personal values), is more significant than the written law. This then underscores why greater consideration needs to be given to ethics.

There is also the question of non-use of the law and EHOs/EHPs who are out in the community may decide against using the law, or enforcing provisions intended to safeguard the health of the populace. How and why are these decisions made? This raises ethical considerations. They are at the very least exercising discretion on behalf of the local authority (assuming that there is a discretionary power). Where there is a legal duty on the local authority to act in the circumstances found but the officer does not act or chooses to use "persuasion" or "informal action" rather than comply with the legal duty, who can then make them, or the local authority comply with that duty? In practice rarely, if ever, is there a comeback in such situations – so it is arguable that this is unethical and is not operating to the highest professional standards.

In general, and in the context of legal standards that are to be enforced, these are often reached as the result of political discussion and arrived at by way of consensus of all vested interests. That is not necessarily the same as the optimum for a healthy environment and public health. It may be complying with the law for an EHO to ensure that standard is met, but is it ethical when they could go further to better safeguard public health? Would it depend on the cost of going further and can the EHO/EHP balance costs against benefit – would there be a marginal gain in health or safety for undue cost? This implies that they can adequately assess risk and benefits. Again, this is something that a fully qualified EHO/EHP should be able to do.

This chapter considers the difference between law and ethics when by the very nature of their work EHOs/EHPs assume they are "ethical" and what they do must be ethical – acting as a friend of the human race. Yet if they see their function merely as regulators or the enforcer of the law is that always the case? Furthermore, what if the law permits or even requires action to protect health but that action is not taken? Then, what if there is conflict between the ethical requirements of their professional body and the law or the interpretation of the law by employers?

Law and ethics not the same?

The law is a means of codifying norms and setting rules that govern society usually set at the governmental level. It has been suggested that while the law "consists of a set of institutionalized rules and regulations that govern society"[4] with the purpose of maintaining some semblance of order. In most countries, the law is developed by the government and in democracies passes into law via a legislative process and is then enforced by government and its agencies and institutions charged with protecting citizens and securing punishment for those who transgress.

As shown already ethics can be said to be more subjective consisting of principles that may be internal to an individual (but may also come from a professional body) that guide that individual's action. Figure 3.1 seeks to indicate the relationship between the law, moral values, and ethics and shows that anything outside of the shapes will be unethical. It can be said that the ethical behaviour of an individual is dictated by that individual's internal moral compass, value system, and personal integrity as well as those that reflect professional or a profession's requirements. As is apparent ethics are the principles that guide an individual or society and help them decide what is right and wrong and indeed how they should behave or conduct themselves. It follows that ethics is something that regulate an individual's behaviour regardless of "the law", even though both law and ethics are intended to regulate or govern human behaviour. For the EHO/EHP this does also imply a need to understand the relevant law (including administrative law)[5] and also be up to date in its application and the views of the courts.

Local authority EHPs who act on behalf of the local authority and with their authorisation, cannot act in any way that is not permitted or required by

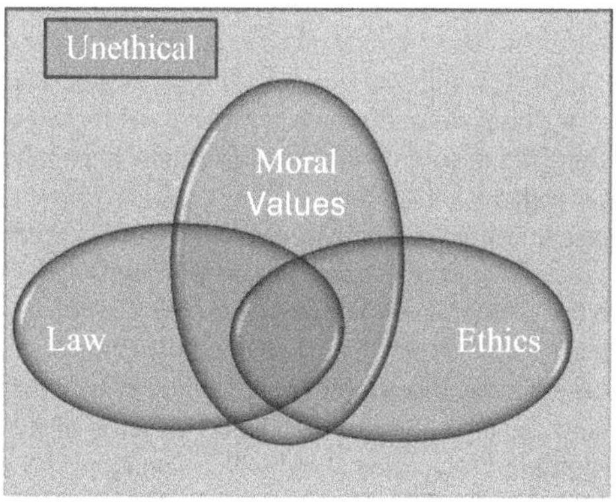

Figure 3.1 Diagram to indicate the relationship between law, ethics, and moral values. This implies that for environmental health practitioners there is a need to keep up to date with the law and how it is interpreted by the courts.

the law. Yet while the law might permit an action and indeed might place a duty on the local authority employing the EHO/EHP, it does not follow that this action or how duty is met, will be met in an "ethical" way. For example, the Environmental Protection Act 1990[6] says it "shall be the duty of every local authority to cause its area to be inspected from time to time to detect any statutory nuisances which ought to be dealt with". The term "time to time" is sufficiently vague as to be open to widely differing interpretations, although total inaction over a long time would be a breach of this duty[7] and a one-off inspection is unlikely be adequate.[8] Yet it is unlikely that this duty is met in a way that improves public health and in most cases action depends on complaint rather than any active seeking-out of nuisances. That said, simply reacting to complaints from the public would not seem to comply with the statutory duty.[8] As McCracken et al. argued "the wording of the section expressly provides for local authority inspections in addition to the investigation of complaints".[8] Although there seems to be no recent Case Law concerning breach of this statutory duty, it would be ethical and wise of an EHP to advise their local authority of the duty and that they should at least monitor and maintain records as to how this duty is met (and that applies to all other statutory duties). It would also be unethical to decide without due consideration of Case Law that something that would constitute a statutory nuisance is not merely for administrative convenience, and this highlights why the issue of ethics is important for a "street-level" officer. From an ethical perspective, this is both a matter of integrity and competence as highlighted in the Chartered Institute of Environmental Health (CIEH) code of ethics and fitness to practise.

At the same time, it is entirely conceivable that the law itself or its implementation might be seen as unethical. What should the EHP do when the law requires them to do something they consider unethical or wrong – surely for a profession "only following orders" is not an acceptable response. Ultimately, EHPs need the right tools to effectively navigate the ethical dilemmas that they may be faced with which can be competing priorities, internal demands, and managerial or political goals which may have a different emphasis from responsible action, and in commercial practice might be perceived as emphasising profit over responsible action.

There is then a question on whether legislation itself is properly drafted or at least drafted clearly so as to achieve its stated objective. For example, in a case decided in late 2024 concerning houses in multiple occupation occupied by asylum seekers[9] with fees paid by the Home Office, the local authority took enforcement action having discovered poor conditions such that they brought 39 charges against the operators for 27 alleged breaches of the Management of Houses in Multiple Occupation (England) Regulations 2006.[10] The problem was the drafting of the Regulations and whether the company providing the accommodation fell within the definition of "manager of the premises" compared with the definition of the "person having control" as in the Housing Act 2004. The High Court found a way to resolve the problem but clearly the law did not help the local authority in their actions to protect the residents of those HMOs. Often legislation follows a failure of the existing legal framework or disaster followed by public concern and review which is influenced by circumstances and the need to respond quickly.

Law, ethics, and the work of EHOs/EHPs (or ethical issues versus policies and practice)

Over the years there has been a debate about the name of the profession, and indeed this book reflects the difficulties. The CIEH talks about EHPs, whereas in Scotland, the term is EHO (in other countries public health inspector is still used). Within local government in the UK the term officer is used. In England as we have seen previously that the law speaks of "proper officer" which could be any professional that the local authority considers to be "the proper officer".[11] Does this matter? Part of the issue and why ethics is important is that increasingly there has been specialisation and the "holistic" approach traditionally taken by environmental health professionals who have been "problem solvers" has led to a narrow legalistic focus. If a problem does not fit neatly into the area of law that person is responsible for, then there is a real risk it can "fall through the cracks" and a member of the community who needs help and support cannot get it. There is also the possibility that someone not qualified as an EHO/EHP can be employed to carry out a limited task, that is, someone who is not bound by professional ethics. This is an important consideration when these people will be out in the community making decisions as "street-level enforcers".

Alternatively, specialisation might result in several different officers being involved. For example, in the areas of food safety and health and safety at work it would be absurd to have one officer visiting a food premises to deal with food safety and another to deal with matters that fall within health and safety at work. In the same way, if a "noise specialist" visits a home on a noise complaint and the rented dwelling is also in disrepair and damp, it does not make sense from the householder's perspective for one problem to be addressed, and the damp ignored or passed to another officer. These are matters that can be addressed under the law, but it is contended that an ethical and even "client-oriented" approach, is for a single EHO/EHP to deal with all issues on the premises, unless it is a particularly complex problem. This requires that the EHO/EHP takes the holistic approach – but that might not fit with what suits managers. This also exposes the limitations of seeing environmental health purely as a law enforcement-led role and also the use of "housing health practitioners" or "food safety practitioners" unless this role is a route to full qualification akin to an apprenticeship.

The question is does "managerialism", expediency, and law intersect here to reduce or even undermine the ethical approach to environmental health? This chapter continues by looking at some issues in the context of the legal framework.

Marketisation

Marketisation can be thought of as something that restructures public services so that organisations such as local authorities seek to operate in a market-oriented way – that is, changing the legal environment in which they operate. This can be done by restructuring management, decentralisation, and privatisation. The aim is to create a system that operates under market conditions.

The influence of money and the lack of resources may have an influence on 'marketisation' and policies and practice that can compromise an ethical approach to environmental health work and a measure of marketisation of that work. At the same there can be financial pressures on non-local government EHOs/EHPs. Somewhere in between are those carrying out environmental health work for local authorities as consultancies or contractors.

Further a non-local government EHO/EHP working for a retailer could be faced with assessing the hygiene where a food product is produced at two plants. Both could comply with the law on food safety at the time of inspection but one has much higher standards, including of management and in which the EHO/EHP has greater confidence than the other, but the product from the second is cheaper. The retailer decides to source the product from the second. Which is the ethical approach? Should the EHP advise against this (or at least provide their professional reservations) or acquiesce on the basis that it is a commercial decision. The CIEH code does not address directly how this situation should be addressed nor give the CIEH view.

Food hygiene ratings

All premises that sell, cook, store or handle, prepare or distribute food should be registered.[12] Registration applies to all types of businesses regardless of where they operate from. It includes businesses trading from home, a mobile unit, or temporary premises such as a stall or van and applies if a business only sells food or drink online. Following a food hygiene inspection, a rating will be uploaded by the local authority so that it is published on the Food Hygiene Ratings website.[13] The theory is potential consumers will avoid poorly rated establishments, so reduce exposure to poor hygiene. Lower exposure should reduce rates of foodborne illness, while simultaneously decreasing the establishment's revenue, applying market forces to improve hygiene. Ethical arguments for posting ratings rest on public transparency as a key leverage point for improving population health. Public health ethics includes preventing harms, maximising and distributing benefits fairly, respecting autonomous choices, disclosing information, and building trust.[14]

A food business will get a new food hygiene rating each time the premises are inspected by the local authority. The frequency of these planned inspections depends on the risk to health from the type of food premises in question, so the greater the risk, the more often it will be inspected. If a business does not receive a rating of "5 – very good", it is possible to request a re-rating inspection to get a new rating before the next planned inspection.

A request for a re-rating inspection can only be requested if the rating has been accepted and once all the necessary hygiene improvements that the local authority food safety officer or EHO recommended at the last planned inspection have been carried out. Some local authorities in England charge a fee to recover costs of a re-rating inspection. This should be notified in the letter providing the rating or when the reinspection is requested. In Wales and Northern Ireland, all local authorities charge a fee for the re-rating inspection under their statutory schemes. While this is of course entirely legal, there is a question as to whether this is ethical and certainly in England where some authorities charge and some do not.

If the local authority does not charge for the re-rating inspection service, it is only possible to have one re-rating inspection between the local authority's planned inspections of the premises. If the local authority in England does charge for the re-rating inspection service, there is no limit to the number of re-rating inspections that may be requested. In Wales and Northern Ireland, the number of requests made for re-rating inspections is not limited, but certain conditions must be met before the local authority agrees to it, including any appeal about the rating having been resolved. The current rating also must be displayed prominently and it also has to be agreed that the inspector will be given access to inspect and re-rate the premises.

Scotland is different with the Food Hygiene Information Scheme[15] that has a different approach from the rest of the UK. It has only three grades, "Pass" for those meeting the legal requirements for food hygiene. "Improvement Required" for those businesses not meeting the legal requirements with improvements needed. Finally, "Exempt Premises" are those that have been inspected and would meet the hygiene requirements but do not meet the criteria to be part of the scheme, for example, low risk and not normally thought of as food businesses such as newsagents and pharmacies. Where improvement is required, a re-visit can be requested before the next planned inspection (inspection frequency reflects risks) and there is no charge for this.[16] There is also the Eat Safe Award scheme for those businesses that go beyond meeting the legal food hygiene standards.[17] The question, regardless of what the law permits or requires, is whether charging for a re-inspection ethical when the main focus should be the safeguarding of public health?

In Northern Ireland and Wales, food businesses are legally required to display their food hygiene rating sticker in a prominent place, for example, at the entrance to their premises. In England, businesses are not legally required to display their rating but are encouraged to do so. It has been suggested that if done properly – with consistent, valid, and reliable measures – mandatory displaying of the rating improves compliance and may reduce foodborne illness. However, compulsory displaying the rating also demands commitment to principles of public health ethics.[14] In England, however, the fact that it is voluntary to display the rating rather undermines one of the purposes – transparency and allowing consumers to make an informed choice.

Revenue raising

Under this heading we consider such matters as licensing fees and civil penalties. This has arisen particularly in relation to housing legislation. The Housing Act 2004 permits the local authority to set its own fees, and these seem to range from £500 to £1,500. For example, for one London Borough, the total cost of the licence is £1,295, payable in two parts, the first to be paid when the application is made of £588.50 and the second to be paid before the licence is issued of £706.50. A metropolitan authority in the midlands has a similar two-part approach, but the fee level depends on the size of the HMO, with (in 2022/2023) a total fee of £1,125 for up to five persons, for 6–11 people £1,215, and £1,300 for 12+ occupants. In another city, the fees depend upon the number of units of accommodation but with seven "sizes" and for 10+ units (the largest category) the first payment is £446.40 (as for all sizes) but the second part is £1,399.60 plus £58 per unit up to a maximum of £3,500. These figures are given to reflect the range of fees that can be charged for mandatory HMO licensing, but the licence period can also vary and can be less than the maximum five years if there are problems. Licence renewals may be charged at a different level. What is interesting is that the compiler of tables of fee levels at the regional level[18] has said that when they embarked on the exercise of compiling the tables, "I had no idea just how complicated the fee structures were and how much they differed from council to council". This raises the question of how fee levels are devised, and is there a chance that the levels go beyond what they should be to raise revenue (even if the law permits the approach) and then is it ethical to take the approach that some councils seem to be taking? Does it take more work to deal with the licence application for a six-unit HMO than a five-unit HMO? It is also true that the law permits the licensing process to be wholly administrative, and no inspection is required before granting the licence. That might not be the best or even ethical approach, but it is permissible. It is assumed that the HMO will be inspected in the licence period (normally five years) but whether or not this always happens is a moot point. In any inspection, the inspector may look for all deficiencies that give rise to hazards, but it seems not to happen. This can be inferred from local authority statistics for England at least. These show that there are an estimated 126,685 mandatory licensable HMOs, with 95,183 of these being licensed but only 2,875 were identified as having a Category 1 hazard under the Housing Health and Safety Rating System (HHSRS) in 2002–2023.[19]

The Housing Act 2004 s.67(4)(a) says LHAs must proceed on the basis "that they should seek to identify, remove or reduce category 1 or category 2 hazards in the house by the exercise of Part 1 functions and not by means of licence conditions". This has the effect of separating the two parts of the legislation and could be why there is no assessment of any hazards using the HHSRS. All that is looked at is what can be done using licensing. So that even if fire precautions are poor such that there would be a Cat 1or Cat 2 hazard, licensing is used as a means to securing upgrading as provided for in SI 2006 No 373, similarly for cold and space heating. A local authority will have "standards" whether good or bad or even justifiable in addition to the SI (which is vague on fire precautions) and that is to what the inspector (who may not be an EHP/EHO) works. Licensing could be seen as a means of generating income and may not require the same level of understanding or knowledge (trained staff) as Part 1 and the HHSRS and there are some authorities that have no qualified EHOs/EHPs dealing with housing.

A study for the National Residential Landlords Association provides some evidence of how civil penalties are used as an alternative to prosecution.[20] The returns from Freedom of Information requests show that the number of civil penalties increased markedly after 2017/2018 but then remained relatively stable afterwards.[21] Some 130 local authorities had used these new powers but most of the total civil penalties are issued by a small group of 20 so-called "super-users" who were responsible for 71% of all civil penalties issued to the private rented sector. The question of ethics and use of civil penalties is raised by the finding that a large proportion of the civil penalties were not focused on attempts to secure improved property conditions but the single biggest reason for issuing a civil penalty was a breach of selective licensing, making up 39% of all reported civil penalties.[22] Civil penalties associated with action to address with hazards and risks to health and safety were relatively low with those relating to conditions or overcrowding accounting for 16% of such penalties.

It can again be inferred that this reflects inadequate funding for local authorities to be able to recruit sufficient staff with the expertise and training to identify hazards, take action, and where there is non-compliance, be able to prove to a criminal standard that a hazard exists, and a notice has been complied with. This touches on the ethical question of whether it is right and beneficial to society to avoid court and raise income rather than ensure the exposure of tenants to hazards is reduced promptly. This is before consideration of where any funds raised go and are used by the local authorities, even if they are supposed to be used to further the local housing authority's statutory functions on enforcement activities in the private rented sector.

While it can be argued that having charging regimes for licences and using civil penalties where there is a failure to comply, such regimes should comply with public health ethics, and whether they improve public health and address health inequity. So far as is known little work has been done to assess the health impact of such regimes.

Air pollution and air quality

Air pollution and air quality can affect our health from before our first breath until our last and is a key public health issue. The World Health Organization (WHO) estimated that in 2019, air pollution caused about 6.7 million deaths.[23]. Of these, almost 85% were attributable to noncommunicable diseases (NCDs), including ischemic heart disease, stroke, lung cancer, asthma, chronic obstructive pulmonary disease (COPD), and diabetes. This makes air pollution the second leading cause of NCDs globally after tobacco.

Air pollution is linked to a range of health problems at every stage of our lives – impacting life expectancy and cutting up to 43,000 lives short a year across the UK[24]. The Royal College of Physicians has suggested around 30,000 deaths per year in the UK are estimated to be attributed to air pollution, with an economic cost of £27 billion in the UK.[25] The impacts of air pollution are felt by everyone but are not felt equally. Although the fashion for using woodburning stoves is an example where people outside the household are affected, the indoor air quality is also compromised.[26] Generally in more deprived areas, residents will more likely be exposed to higher levels of pollution, and it seems people from ethnic minority communities are much more likely to be exposed to toxic air. Certain groups are also hit harder when it comes to health. Children, older adults, and those with existing health conditions are especially vulnerable to its harmful effects.[27] Thus, it is a health equity (and ethical) issue. In 2020, the coroner found that air pollution was a factor in the death of Ella Adoo-Kissi.[28]

There are various legislative provisions relating to outdoor pollution. For indoor air pollution, there are hazards in the HHSRS that consider defects that could cause problems of indoor air quality that can be addressed using the Housing Act 2004. Some 3% of global deaths have been attributed to indoor air pollution.[29]

The Environment Act 1995, as amended, provides the basis for local authorities' obligations to monitor and report on air quality in their areas, usually an environmental health function, and take appropriate action including the use of Air Quality Management Areas (AQMAs). An AQMA is an area where local air quality is unlikely to meet the government's national air quality objectives. Under Section 83(1) of the Environment Act 1995, local authorities are required to designate areas where air quality objectives are not being met. The extent to which local authorities see air pollution as a priority is variable, yet EHOs/EHPs should be arguing it is a public health priority.

The Environmental Permitting (England and Wales) Regulations 2016, as amended (EPR 2016) set standards and provisions to reduce the emissions of pollutants from a diverse range of industrial sources, usually the smaller industrial sources. These sources often emit less visible pollutants.

The Clean Air Act 1993 deals with more visible pollutants and provides for a power for local authorities to designate smoke control areas, but it is not a duty. There is no obligation to designate such areas, but where there are issues, EHOs/EHPs should be recommending such declarations. The Environment Act 2021 allows local authorities to issue financial penalties for emitting smoke from a chimney in a smoke control area in England. It means that in England, emitting smoke from a chimney is no longer a criminal offence, but rather subject to a civil penalty notice (a fine). The aim of this change was to make enforcement "quicker, simpler and more proportionate". The Environment Act 2021 amended the Environmental Protection Act 1990 so that smoke emitted from a private dwelling in a smoke control area in England could be defined as a statutory nuisance if it were "prejudicial to health or a nuisance".

There is no shortage of legislation to secure improvements in air quality and how local authority EHPs/EHOs use this is a matter of health equity and indeed what use is made of the law, is a matter of ethics. Whether the air quality standards set by the government are adequate or sufficiently challenging to prevent ill health is another question. Air pollution is clearly a matter of public health. On that basis EHOs/EHPs should be working closely with their Directors of Public Health[30,31], planners and other agencies and even schools as some EHPs have shown[32] to monitor air quality, ensuring best use is made of the legal framework. If the standards are considered inadequate to protect public health, then they should work with their and other professional bodies to get these changed.

Awaab Ishak and the damp and mouldy flat

Reading the Coroner's report to prevent future deaths, the first reaction was where were the EHOs/EHPs and why had they not inspected the property? Reports and photographs of the conditions in the press indicated there was most likely a Category 1 hazard for Damp and Mould. Local housing authorities have a duty to take one of the courses of action in Part 1 of Housing Act 2004 in such circumstances. It seems, however, that the environmental health department had not been contacted. The property where Awaab lived is owned by Rochdale Boroughwide Housing Association (RBH). It is not known whether there was a policy of environmental health not inspecting housing association properties but the published enforcement policy refers only to privately rented accommodation.[33] It might have been that there was a lack of understanding of the role of EHOs in assessing housing conditions and taking action on the part of the Health Visitor, general practitioner (GP), or legal advisor, who did not consider contacting the environmental health department. Neither the coroner nor the housing ombudsman[34] referred to the role of EHOs in such cases. The "invisibility" issue again.

The coroner was satisfied from the evidence that the mould remained a continuing and recurrent issue while the family were in the property. In 2019, the family made an application to be re-housed. In July 2020, the family showed a Health Visitor the mould and a letter was sent dated 9 July 2020 to RBH by the health visitor explaining their concerns about the mould and potential impact on Awaab's health – did the health visitor know that the EHO could take legal action to secure a remedy, and if not why not? Throughout his life Awaab had recurring cold symptoms such as runny nose, cough, and respiratory tract infections. His GP confirmed that he attended at their surgery more than most children.

Solicitors were instructed in 2020 to make a disrepair claim – it was entirely possible to use the Landlord and Tenant Act 1985 as amended by the Homes (Fitness for Human Habitation) Act 2018. While RBH visited the property and saw its condition, the policy at the time was not to progress the remedial work until agreement with the solicitors. This is entirely wrong. As Giles Peaker pointed out[35] the coroner highlighted a "policy" amongst housing associations, in cases where a disrepair claim has been brought, of waiting for agreement from the claimant (or their legal representative) before rectifying any recognised disrepair. He said, "It has no basis in law or in the Housing Conditions Pre-Action Protocol…..it just needs landlords (and some claimant solicitors) to acknowledge the stupidity of this, stop messing about, act according to the law and follow the pre-action protocol". One wonders if there was also a policy in the local housing authority of not inspecting housing association properties, just as there is (or has been in the past) a policy of not inspecting local authority-owned properties?

The coroner's report said that it was only following Awaab's death the local authority carried out an HHSRS report with a view to enforcing any remedial action required. One issue for the environmental health profession is why three members of other professions involved in this case did not think to contact the EHOs about the conditions. If it should be that (and not unusually) there is a policy of not inspecting or responding to complaints about conditions in social housing, then that would pose an ethical dilemma for the EHO. Housing associations are not exempt from Part 1 of the Housing Act 2004 (even if they are from licensing provisions in other Parts). It is arguable that failing to inspect is not only a breach of a statutory duty, it is unethical as it is discriminating against one part of the community which is entitled to the protection afforded by EHOs/EHPs. This then should be where the professional body has a role in highlighting a policy that runs counter to ethical expectations. Defending the "status quo" or acceptance should not be the first approach even if there is some sympathy for lack of resources.

Grenfell Tower

That the government saw fit to publish an addendum to the HHSRS Operating Guidance dealing with fire in high-rise buildings indicated that

EHOs/EHPs had a role in reducing risks and could act to improve fire safety in common parts and individual flats in such buildings. However, the Report of the Grenfell Tower made no mention of the role of the Borough's environmental health department, focussing not surprisingly on the Building Control department and on the poor relationship between the Tenant Management Organisation (TMO) and the Borough and between the residents and the TMO. Within Grenfell Tower, however, there was a mix of tenures. On 14 June 2017, there were 14 leaseholders of flats within the tower; the remaining flats were home to social housing tenants. The vast majority of the residential flats in the tower were part of the Royal Borough of Kensington and Chelseas's provision of social housing within the borough.

The Grenfell Tower Inquiry in Part 1[36] highlighted doors lacking self-closers or ineffective self-closers. For any of those flats owned by leaseholders, the EHO could have acted using powers in the Housing Act 2004, and this is a deficiency that a competent EHO/EHP would have identified.

As well as again raising questions about the invisibility of environmental health, there is also a question as to whether or not the local authority's EHOs were even permitted to inspect.

Surface water pollution – sewage

The quality of surface waters is a continuing issue as the result of sewage pollution and agricultural run-off. EHOs have the powers to ensure that near house drainage is adequate and working correctly even if they have little power to ensure public sewers are working correctly and some watercourses are not polluted.[37]

The question sometimes arises as to what the EHO can or should do about sewage pollution of watercourses by sewage discharges including where detritus is deposited on land (riverbanks) given the risks to health in such circumstances for wild water swimmers and others who use the waterways and their banks for leisure activities. There are two instances where the law including Case Law would support action by local authority EHO/EHPs to take action to protect public health, for example, sewage litter on riverbanks is analogous to sewage on a beach, although care needs to be taken in how this is done.[38,39] Where sewage discharges onto land as the result of a defect in the sewerage system (even if operated by the water and sewerage company (WASC)) action can be taken if those premises are in such a state as to be prejudicial to health as a consequence.[40] Although in the case referenced here it was not liable as the source of the sewage was not coming from a public sewer (the responsibility of the WASC). The ethical question given the public concern is why the law is not being used more by local authorities and their EHOs on this issue.

Charging for local authority services

Using discretionary fees and charges is one of the key income-generating options available to local authorities. Section 93 of the 2003 Local Government Act allows authorities to charge for services that they have a power (but not a duty) to provide. While the income they can earn from these charges is restricted to the cost of providing them, there is no restriction on how the costs are calculated. While fees and charges cannot be used to make a profit, they could provide the opportunity to invest in infrastructure, because all aspects of service provision can be included in cost. Authorities can also use differential charging based on people's ability to pay.[41]

Charging is possible for services in addition to those funded from Council Tax and central government grants. These include services such as leisure centres and parking and can only offset the cost of these services. Councils can also generate income from sales, for example, of property, and other charges, such as for certain types of waste removal. Local authorities increased the money raised from sales, fees, and charges by £1.6bn between 2010/2011 and 2019/2020. In 2019/2020, service users directly funded almost half the cost of planning and development services (e.g., planning fees) up from 25% in 2010/2011. Direct funding for highways and transport services (such as fees for supervising works) increased 31% between 2010/2011 and 2019/2020.[42] The financial pressures on local councils will increase the chance of such charges increasing. While not a direct environmental health issue there should be an assessment of the impact on public health and health inequity of such charging regimes.

Infectious diseases

Under the Local Government Act 1972, local authorities at upper tier, unitary, and lower tier levels have a statutory duty to provide a range of services to their communities in relation to outbreak management. While there may be some EHPs at the upper tier, most are in the unitary and lower tier. Contact tracing is normally undertaken by UK Health Security Agency (UKHSA) Health Protection Teams (HPTs). However, in some circumstances, others may undertake or support contact tracing work, including local authority EHOs.[43]

An outbreak is defined as either:

- an incident in which at least two or more people affected by the same infectious disease are linked by time, place, or common exposure
- a greater than expected rate of infection compared with the usual background rate for the place and time where the outbreak has occurred

Other incidents that can be considered in a similar way to outbreaks include:

- a single case of certain rare or high-consequence diseases, including, but not limited to, diphtheria, botulism, rabies, viral haemorrhagic fever, and polio
- a suspected, anticipated, or actual event involving microbial contamination, for example, food, water, and medicine or medical device contamination

Reporting of notifiable diseases is outlined in the Public Health (Control of Disease) Act 1984, the Health Protection (Notification) Regulations 2010[44], and also the Health Protection (Notification) Regulations 2025.[45]

Environmental investigations include food chain analysis and investigations, and, in some circumstances, there can be a need to characterise risk of harm from environmental sources of infection and determine if control measures focused on the environment or food chain are required. This is where EHPs/EHOs are involved and can play a key role in environmental investigation, including the collection of samples to support microbiological investigation where required.

EHOs/EHPs should be in a position to use their skills to help determine the causes of disease outbreaks, prevent the further spread of illness in their communities, and provide information and advice to patients on how to minimise infection in their homes. It is clear that EHOs/EHPs have to behave ethically in such circumstances.

Contaminated land

In 2025, Netflix aired a programme called Toxic Town very much based on the litigation surrounding contaminated land and how Corby Council dealt with decontamination in the 1990s. Contaminated land had been raised as a political and environmental health issue in the 1980s as the result of the closure of many heavy and traditional industries. Much of how contaminated land should be dealt with is included in a chapter by an EHP in the 22nd Edition of *Clay's Handbook*.[46] He acted as an independent consultant in litigation where it was alleged Corby Council carried out remediation work on a disused steel works site negligently. In *Corby Group Litigation v Corby Council*[47], it was claimed that, between 1986 and 1999 some children were born with birth defects as a result of negligence, breach of statutory duty, and public nuisance on the part of the Council. It was claimed that the birth defects were due to expectant mothers inhaling or ingesting harmful substances (cadmium, chromium, nickel, dioxins, and polycyclic aromatic hydrocarbons, on their own or in combination) generated by the reclamation.

The Court found in favour of the claimants, accepting the expert opinion that those responsible for the reclamation works had seen the project as simply an "engineering" operation involving often unspecified material, rather than as a highly complex remediation and reclamation process. It ruled that the Council had been "extensively negligent" in its control and management of the reclamation of the land, and held the Council liable in public nuisance, negligence, and breach of statutory duty. The Court ruled that these breaches had led to the extensive dispersal of contaminated mud and dust over public areas of Corby and into private homes, a consequence of which were birth defects in children whose mothers had inhaled the contaminants.

This case is included here as it raises some questions as to the role of the local authorities' EHOs and what advice they provided – or even whether they were consulted. What would they have done if they had assessed the situation and had come to the same conclusions as the EHP (Roger Braithwaite) whose evidence was referred to in the judgement? Where would they have gone or what could they have done. This ought to be a role for the professional body.

The pressures on local councils to bring in business and economic activity are no less now, so the argument is that professional bodies need to have a robust system in place for providing support, including a whistleblower policy. In discussions with colleagues on the matter of professional ethics, it was suggested that the CIEH would be the last place to which they would turn to in such a situation – this is unsatisfactory and should not be the case.

"Whistleblowing" law

UK law has granted protection for those who raise concerns of potential wrongdoing in the workplace (whistleblowers) since 1999 through the Public Interest Disclosure Act 1998. Legally this is known as "making a disclosure in the public interest" but is best understood as "whistle blowing".

A whistleblower is protected in law if they reasonably believe that the wrongdoing they disclose is in the public interest – this depends on what is called the Chesterton test.[48] This considers the number of people affected, the nature and impact of the wrongdoing, and who the wrongdoer is, and the wrongdoing disclosed must be in the public interest so must affect others such as the general public.

It seems to the authors that EHPs are by the nature of their work very much in a position that could justify whistleblowing, so this is something that their professional bodies should consider how they address this and provide advice and support to members of the profession who wish to raise concerns.

Conclusions

The law and ethics are not the same thing and from an environmental health perspective it is whether the law is being used ethically. Conversely and perhaps more importantly is failure to use the legal powers available to resolve an environmental health problem or to protect public health, ethical. Or could it be that it in certain circumstances use of environmental health law is considered inappropriate for the problem, in which case there is a need for further guidance.

As so many EHPs/EHOs consider that their primary role is as a regulator, yet actual enforcement is relatively rare. It is thus important that they understand the difference between the law and ethics, and this is something that the professional bodies could usefully address. It perhaps goes beyond what is "professional ethics".

Notes and References

1 https://www.cipfa.org/cipfa-thinks/articles/fees-and-charges-a-significant-income-for-councils

2 Hutter BM, (1988) *The Reasonable Arm of the Law? The Law Enforcement Procedures of Environmental Health Officers*, Clarendon Press, Oxford.

3 Hutter, BM, (1989). Variations in Regulatory Enforcement Styles. *Law &Policy*, Vol. 11, No. 2, pp.153–174.

4 https://www.ethicsatkentplace.org/about-ethics/ethics-institute-blog/individual-blog-page/~board/blog-posts/post/the-relationship-between-ethics-and-the-law

5 Administrative law in essence is the relating to the control of the power of government, its agencies and public authorities including local government see Wade W and Forsyth CF, (2014) *Administrative Law* 11th Edition, OUP, Oxford.

6 At section 79(1).

7 See *Meade v LB Haringey* [1979] 1 WLR 637.

8 See Interpretation Act 1978 s 12 and McCracken et al. (2001) *Statutory Nuisance*, Butterworths, London.

9 *Clearsprings Ready Homes Limited, R (on the application of) v Swindon Magistrates' Court No 2* [2024] EWHC 3245 (Admin).

10 SI 2006 No 372.

11 Section 112 of Local Government Act 1972 says a "local authority shall appoint such officers as they think necessary for the proper discharge by the authority of such of their ......functions as fall to be discharged by them".

12 The Food Premises (Registration) Regulations 1991(SI 1991 No2825) made under the Food Safety Act 1990 require that in general "no person shall use any premises for the purposes of a food business on five or more days, whether consecutive or not, in any period of five consecutive weeks unless those premises are registered".

13 https://ratings.food.gov.uk

14 Fleetwood J, (2019) Scores on Doors: Restaurant Hygiene Rating and Public Health Policy. *Journal of Public Health Policy*, Vol. 40, pp. 410–422. https://doi.org/10.1057/s41271-019-00183-4

15 https://www.foodstandards.gov.scot/consumers/food-safety/buying-food-eating-out/food-hygiene-information-scheme/about-the-food-hygiene-information-scheme

16 https://www.foodstandards.gov.scot/publications-and-research/publications/fhis-safeguards

17 It should be noted that The Food Hygiene Information Scheme and the Eat Safe Award run side by side. The Food Hygiene Information Scheme rates food outlets

against the legal food hygiene requirements, while the Eat Safe Award recognises food businesses that strive for excellence in delivering Food Hygiene standards.

18 Tables of licence fees at the regional level were published at https://theindependent-landlord.com/licensing-fees/

19 https://www.gov.uk/government/statistical-data-sets/local-authority-housing-statistics-data-returns-for-2022-to-2023

20 Civil penalties were introduced by the Housing and Planning Act 2016, and guidance can be found at https://assets.publishing.service.gov.uk/media/5ac63c47ed915d76a313cbd5/Civil_penalty_guidance.pdf

21 Wood J and Watkin S, (2021) The Enforcement Lottery: Civil Penalty Usage by Local Authorities. *Working Paper Series #5.* https://www.nrla.org.uk/research/special-reports/enforcement-civil-penalty-usage

22 This is the only type of civil penalty that is unrelated to attempts to improve property conditions.

23 https://www.who.int/news/item/25-06-2024-what-are-health-consequences-of-air-pollution-on-populations

24 UKHSA (2023) Health Effects of Climate Change (HECC) in the UK: 2023 Report, Crown copyright. Available at https://assets.publishing.service.gov.uk/media/6570a68b7469300012488948/HECC-report-2023-chapter-4-outdoor-air-quality.pdf

25 RCP (2025) *A Breath of Fresh Air: Responding to the Health Challenges of Modern Air Pollution.* Available at https://www.rcp.ac.uk/policy-and-campaigns/policy-documents/a-breath-of-fresh-air-responding-to-the-health-challenges-of-modern-air-pollution/

26 See https://www.sheffield.ac.uk/news/indoor-wood-stoves-release-harmful-emissions-our-homes-study-finds

27 https://ukhsa.blog.gov.uk/2018/11/14/health-matters-air-pollution-sources-impacts-and-actions/

28 See https://www.bmj.com/content/371/bmj.m4902

29 See Hands A, (2022) Air Quality, in Battersby S, Ed, *Clay's Handbook of Environmental Health*, Routledge, Abingdon, Oxon, pp. 700–747.

30 See https://laqm.defra.gov.uk/assets/63091defraairqualityguide9web.pdf

31 See, for example, South Gloucester Director of Public Health Report 2022: *Clean Air and Climate Change.* https://beta.southglos.gov.uk/static/aac40bf-4098999caa82790f413760837/DPH-report-2022.pdfandNorthTynesideDirector of Public Health (2023) *Health Protection Assurance Report.* North Tyneside Council.

32 Taylor P, (2025) Clean Air Champions. *Environmental Health News*, June 2025, pp. 10–12.

33 https://www.rochdale.gov.uk/downloads/file/1350/private-sector-housing-enforcement-policy-document

34 Housing Ombudsman Service (2023) *Housing Ombudsman Special Report on Rochdale Boroughwide Housing.* https://www.housing-ombudsman.org.uk/wp-content/uploads/2023/03/P49-RBH-FINAL-200323.pdf

35 https://nearlylegal.co.uk/2022/11/awaab-ishak-coroners-ruling-and-reg-28-report-to-prevent-future-deaths/

36 https://www.grenfelltowerinquiry.org.uk/phase-1-report

37 See the Building Act 1984 s.59, for example, and Public Health Act 1961 s.17 and provisions in the Public Health Act 1936 including nuisances in connection with watercourses and ditches.

38 *R v Carrick District Council, ex parte Shelley* [1996] EWHC J0403-1.

39 *R v Falmouth and Truro Port Health Authority, ex parte South West Water Limited* [2000] EWCA Civ J0330-10.

40 *City of Bradford Metropolitan District Council v Yorkshire Water Services Ltd* [2001] EWHC Admin 687 [2002] 1 EGLR 85.

41 https://www.cipfa.org/cipfa-thinks/articles/fees-and-charges-a-significant-income-for-councils

42 https://lordslibrary.parliament.uk/local-government-finances-impact-on-communities/#heading-6

43 https://www.gov.uk/government/publications/communicable-disease-outbreak-management-guidance/communicable-disease-outbreak-management-guidance-principles-to-support-local-health-protection-systems

44 SI 2010 No. 659.

45 SI 2025 No. 291.

46 Braithwaite R, (2022) Contaminated Land and Land Use, in Battersby S (Ed), *Clay's Handbook of Environmental Health*, Chapter 19, Routledge, Abingdon, Oxon, pp.855–911.

47 *Corby Group Litigation Claimants v Corby Borough Council* [2009] EWHC 1944 (TCC). Available at https://www.bailii.org/ew/cases/EWHC/TCC/2009/1944.html

48 *Chesterton Global Ltd and another v Nurmohamed* [2017] IRLR 837 CA - The Court of Appeal upheld the decision of the employment tribunal that disclosures made by a worker satisfied the "public interest" requirement for protection under the whistleblowing provisions of the Employment Rights Act 1996. The disclosures related to a breach of the employment contracts of 100 senior managers, including the whistleblower.

4 Ethical dilemmas and concerns and resolving these (or not) – individual testimonies and comments

This chapter considers different cases, scenarios, and recollections from environmental health officers (or practitioners) (EHOs/EHPs) concerning ethical questions and issues they have faced or experienced in their careers and how these have been resolved or if they have not what the outcome was in the end. It is drawn from interviews or voluntary submissions from EHOs/EHPs who were contacted either directly or via a request for feedback and thoughts on LinkedIn. These dilemmas reflect some of the issues considered in previous chapters, reinforcing some of the points made previously or raising new ethical questions.

These have been anonymised, and prior to publication all the contributors have had the chance to comment on what has been included here. Where they are testimonies and direct quotes the text has been boxed.

Scenarios and testaments from EHOs/EHPs

Practitioner 1

In this instance, the EHO employed by one local authority (but now retired) was faced with a dilemma regarding his grandmother and the failure of another local authority to deal adequately with conditions in her home and how this should be handled. It raises the issue of whether it is ethical for an EHO not to inspect dwellings provided by the employing local authority. In this case, the inspection by the local authority EHO was only undertaken at the request of the Chief Executive after the intervention of the tenant's grandson, Practitioner 1.

He had complained to the Local Government Ombudsman (LGO), but in the Ombudsman's Report their name was changed.[1] The complaint, made on behalf of his late grandmother who had been blind, alleged that a London Borough had failed to deal properly with the housing benefit claim, a disabled facilities grant application, and repairs to a damp wall at the rear of her property owned by the borough. As the result of the failures at the local authority,

DOI: 10.1201/9781003625339-4

he had instructed and paid the fees for a private Disability Living Consultant and a private surveyor. The LGO report sets out the EHO's allegation that the property was probably unfit for habitation under the relevant housing law. The EHO grandson had made representations to the Chief Executive asking what steps the Council were taking to fulfil their legal obligations as landlords. The Chief Executive consulted an EHO of the authority. That EHO also confirmed the view that some of the defects rendered the property "unfit" under the terms of the Housing Act applicable at the time. There has never been a reason why an EHO could not inspect and report on conditions in a property owned by their employing authority, it is just that they cannot take any legal action. At one time there was even circular advice that such inspections should be undertaken.[2,3]

The Council had then failed to ensure that essential works were completed. Practitioner 1 subsequently had to instruct a private surveyor to provide a review of the situation, as they had chosen not to carry out an inspection themselves.

The LGO made a finding of maladministration with respect to the conditions in the home and the DFG (but not the HB claim). The Council's delay in dealing with the DFG amounted to maladministration. Ultimately, however, the decision was that she was not eligible. The delay did cause her injustice, and it took over two years after the DFG application for two additional grab rails to be fitted. The landlord local authority was made aware of dampness in the flat in 1988. There was unreasonable delay in repairing gutters (at least four years); rewiring (nine months), repair works to the roof were ordered in July but were not carried out until November 1991, renewing rotted timber in windows took 20 months. It had taken the intervention of the private surveyor in 1992 for many works of repair to be ordered. The Borough's own EHO had prepared a detailed schedule of defects in December 1990.

This EHP also had an ethical decision to make following the birth of his daughter as there were concerns that in the maternity unit there was a failure concerning the best way to support mothers and secure the care and conditions they need. This EHP decided to publicise the problems experienced in local maternity services. The support of the local press enabled the much-needed improvements to be expedited.

Practitioner 2

This case reflects the experience of an EHP who at the time of writing was employed by a local authority in a senior position but had experience of working outside local government. This EHP had found that colleagues could take a position on housing conditions of defending the council's actions or inaction even when the evidence suggested (and the EHO knew or should have known) that there was a problem within the council. In this instance, the onus was then placed on the tenant to take their own action when potentially the council via the housing service was at fault.

It was further exemplified in circumstances when there is legal action and the local authority EHO is called by the defence (the local authority) to give expert evidence, given their supposed expertise on housing conditions. It is possible for this to be permitted but the expert witness has a duty to the court. The experience of Practitioner 2 is they do not fully understand that duty and for whatever reason they are prepared to try to "defend the indefensible". The court will or should discount such evidence if the EHO does not demonstrate that they understand this duty.[4]

That colleague in submitting their views to the authors raised the issue of whether this attitude reflects inadequate training or whether it is something to do with the EHP not being a protected profession (as a statutory appointment). It was also suggested that "the professional body no longer has the reach that it once did into local authorities". Also (and this seems contrary to comments from the Chartered Institute of Environmental Health [CIEH]) many local authorities are not prepared to require EHPs to be members of the CIEH. As the EHP says, this then

> *begs the question – how can the CIEH control ethics and provide direction when EHPs are not members and in such circumstance, the only recourse would be to the council's own whistle blowing policy or formal complaints process when the person investigating would be using council processes.*

The professional standards laid down by the CIEH then become irrelevant and as this colleague says "hence, I would never consider recourse to the CIEH to resolve/address ethical issues or even incompetence".

This colleague also commented that they had experienced conflicts with independent environmental health experts who were prepared to be less than truthful in court in housing cases but in such circumstances would not have considered the CIEH as the place to go to complain. The hope was that they would be discredited in court and viewed as an unreliable witness. It was also suggested that some independent "EHPs" were advising landlords groups in actions against councils but were wholly unprofessional and with no apparent "ethics". Again, the CIEH was not seen as relevant to these concerns. By comparison it was said that if there was any similar concern with a Registered Building Control Inspector, then a complaint would be made to the Building Safety Regulator.

This EHP also reported on the conflicts that can result between planning and environmental health when planning consent is being sought in areas of poor air quality or poor land quality when the push from the council is to increase industry and therefore revenue coming into the council. This can result in "pressure being put on environmental health to agree to planning applications when that is contrary to public health". While not directly affected this colleague saw pressures developing now around EH/housing and the need for temporary accommodation with poor condition accommodation being used even when vulnerable adults may be put at risk, even by adult social care who "have the mandate to protect them".

Practitioner 3

This EHP reported that in relation to a contaminated land scenario, where advice to the council – including the senior management – was that substantial work needed to be undertaken to address an area of significant contamination, but that would be expensive. Ultimately the advice was not taken, and potentially public health was impacted by the limited remediation undertaken. In such circumstances, although the EHP's priority was to public health, the senior leadership overruled the action believed to be necessary for public health. The dilemma was where to turn and how is it possible to get accountability in such circumstances and ensure that the EHP's advice is on public record? It was suggested that there were similarities with the scenario highlighted in the Corby case and the "Toxic Town" Netflix series referred to in Chapter 3. The concern was that there was serious industrial contamination with a potential threat to neighbouring residents and potential residents of the site as there was a desire to sell the land for residential homes. Again, there was no reference to the CIEH, and indeed, "I think an officer would be roundly criticised, even disciplined if they did so – it would be seen as being an internal issue and to be dealt with as such". This situation underscores the argument for some statutory backing for EHOs/EHPs.

Practitioner 4

By way of introduction this is a thoughtful testimony that is the reflection of an EHP who eventually left local government. Again, it is thought provoking and raises issues, but the question of ethics and ethical behaviour has been considered in some depth. The practitioner wrote the following in the boxed paragraphs.

> The most difficult conversation about ethics I had was not with any leadership team but many years after the events was with myself. Throughout my career it was believed that serving the needs of my employer was an expected professional behavioural value system. This was a significant factor in how to navigate the complexities of my professional judgements and decision making. My employer required compliance in all cases but did not invest in capacity for enforcement nor had a positive learning culture, most problems were deemed a very big problem and the first question asked was often "who?" and not "why or how?".

Where every effort was made to make peoples' lives safer, ensuring broad compliance where I could as it was not realistic to achieve total compliance. The approach was a coping mechanism in a challenging work environment and dysfunctional organisational culture. This approach was not shared with my manager but was an approach silently adopted by most of my peers. This reflects what Lipsky[5] described how the street-level bureaucrat works. I had previously enforced, and tacitly supported decisions known to be unreasonable, but these had been directed by my employer as perceived to be in the "public interest".

On reflection they were in the organisation's interest, but these decisions went unchallenged so now I was complicit with my silence, one motivation was career preservation and a desire to progress and to move up higher within the local government organisation. It has been argued that an EHO, like any regulatory official, undertakes a multiplicity of roles and activities from the pragmatic to the symbolic. This leaves a scope for different approaches and silence can be justified to oneself[6] and also officers remain accountable to and under the control of the council.[7] So the political decision-making that undermined any legitimate enforcement actions within the local authority directorates was rarely challenged. For example, this could be not prosecuting or delaying enforcement action or agreeing to damaging press embargos for corporate clients who had other more significant relationships with more influential parts of the local government machine.

I knew that this did not align with the principles that had drawn me to the environmental health profession in the first place, but I convinced myself this was a matter of pragmatism, this was local government where idealism had no place. A question of "don't rock the boat or plough your own furrow" and "follow your last order" were maxims often shared by colleagues to try to navigate these ethical dilemmas.

It took some years after and after leaving the organisation, to fully reflect on the compromises that had been made and my participation in a system that often prioritised self-interest over public good. That said with colleagues we had improved some peoples' lives but perhaps we could have made an even greater contribution to society as EHOs.

The second area where an ethical question arose when as a manager pressure had been put on colleagues to meet unrealistic and unreasonable targets, knowing that the management system itself was dysfunctional and the consequences unfair to EHOs in the teams. "The unspecified threat of dire consequence" had been communicated by proxy there was complicity in a micromanagement bullying blame culture where judgement-free learning, taking risks, and innovation were not priorities. This was a political arena, and the decision-making model was not one based on intellectual debate or collaboration.

It was never imagined that my career in environmental health would be navigating so many ethical dilemmas. When as a manager, there was a vision of implementing social regulation with integrity and fairness but the reality of working within a bureaucratic siloed organisation slowly eroded my idealism. Decisions were not always made based on public interest or ethical correctness; they were often dictated by political expediency, risk aversion, and the unwritten rules of a deeply entrenched culture of self-preservation.

The pressure to conform in ways that conflicted with my morals was frequent, when it was required to maintain an exhaustive performance management spreadsheet tracking EHO non-compliance with management instructions on their every single case. This was an exercise less about accountability and "more about blame and not a collegiate team approach to public health". On at least two occasions there was a direction to enforce policies that negatively affected vulnerable groups while shielding those with corporate or political clout. It was also true that I contributed negatively to employees' stress and well-being.

When the courage to challenge and to speak truth to power was found, the consequences were predictable. My views were dismissed, motivations questioned, and it was suggested that such representation was an ill-considered career move. For years, the fear of retaliation had been internalised and held within the weight of systemic dysfunction. With external support fears were confronted and the culture of micromanagement, bullying and intimidation called out. Staying silent was at the cost of job satisfaction and sense of purpose but speaking out was both a professional decision and one which improved both personal well-being and growth.

Ethical dilemmas are not unique to the environmental health profession but were frequently observed in environmental health practice. Colleagues struggled with the same conflicts between what was right and what was expected, others navigated this more courageously. Some had the credibility to stand up to the managers, quiet acts of defiance that preserved their integrity. Others, like me, adapted, justifying their actions in the name of job security and professional survival.

Other colleagues engaged in practices that blurred ethical lines; even bending policies to suit political pressures and turning regulatory work into private consulting opportunities. There was also a sense of disillusionment, with a creeping realisation that they were part of a system that did not always serve the people it was designed to protect, most eventually left and thrived in pastures new.

When speaking openly about these challenges, there was a mix of relief and resistance. Some colleagues admitted to similar frustrations, grateful for the chance to share their own struggles while others shrugged resignedly.

Throughout these moments of ethical crisis, there was never a question of from whom could help be sought? Whistleblowing carried enormous risks, often leaving those who spoke out ostracised. There was no knowledge of any environmental health teams or organisations that had meaningful frameworks for ethical guidance – only high stakes public grievance procedures. These were designed more for organisational damage control than accountability and progressive change. Were there really any options? There might have been such as through the professional body or trade union "but I never reached out and internalised my doubts and carried the weight of compromise alone". Self-imposed isolation making it even harder to challenge the status quo.

The turning point came not in a big bang but in quiet reflection and the realisation that the impact of some my work had not had the impact for good but even in some cases had crossed an ethical line. Several hoarding cases were haunting; instances where enforcement action led to the displacement and demise of vulnerable individuals, "where regulatory processes ignored the human cost of compliance". I had at times been part of a system that prioritised regulation of risk over people's well-being.

Leaving local government was to my mind the first real ethical decision made in years. In doing so, there was a new path that allowed me to advocate for change, to use my experiences to inform and reform practice, rather than perpetuate the local government status quo.

"In hindsight should I have spoken up sooner, should I have left earlier?" Doing the right thing is not just about recognising right from wrong; it's about having the courage to act despite the risks. The career in environmental health is not regretted but the years spent in silence are, the times when compliance was chosen over conscience. It is hoped that others finding themselves in a similar position will have somewhere to turn for support and advice and will not wait as long to speak their truth.

Ethical practice in environmental health is not about the choices we make in isolation. It is about the collective commitment to integrity, the willingness to stand up, speak out, and demand better and perhaps speak truth to power.

Practitioner 5

This testament is from an EHP who has worked in local government, academia, and the private sector, and who qualified in the 1970s. It records the ethical issues faced when the AIDS/HIV crisis arose in the late 1980s, and environmental health assumed a novel public health role.

Our now retired but still active, practitioner recalls that the Diploma in Environmental Health taught at Tottenham Technical College contained no explicit reference to ethics or the principles of ethical practice. However, he recalls, what the part-time course did provide was continuous classroom and work-based instruction by practitioners and former-practitioners over four years in the "art" of effective environmental health practice. Along the way, the teaching made clear there was a "right" and "wrong" way of achieving the desired outcome. Although there was a joke about applying the "The Ways and Means Act", Practitioner 5 believes the values of ethical behaviour were instilled in his cohort of "pupil public health inspectors", whether in describing the limitations of the officer's powers, or the need to consider the rights of the individual to expect their interests and confidentiality to be honoured and respected.

Here, in his own words, is the story that he believed shaped his career after qualification. Again, the quoted text is boxed.

> As I made the transition from student to qualified officer, I saw the consequences of neglecting the rights of those with whom I, and my colleagues, had dealings. There were some rituals that I look back on now as important for routinely and regularly reinforcing the need to act properly and not just within the law. One, the annual reaffirming of the EHPs enforcement powers and duties by the Council, saw us reissued with a 'warrant/identification card'. This may have been no more than formality in some places, but my chief officer at the time had come from a local authority that had ridden rough-shod over process and procedure, and this was not going to happen under his management.
>
> While I might have used slightly dubious means of securing access to a premises or property in my first post, the experience in my new authority of defending the procedural record of my predecessors, left a lasting impression. The case involved a long-running domestic nuisance case involving cats, and although a generation of EHOs had cause to make visits, interview complainants and serve notices, the LGO was as interested in knowing how information was secured and shared, and how the different parties were treated, as he was in the outcome.
>
> The only other encounter with situations that fell well short of the ethical standard expected, concerned stories shared with colleagues

from other authorities, that certain privileges and advantages seemed to be granted to some individuals because of their "association" with council members or senior officers. I have to say my own superiors saw influence of this kind as reason to pursue robust action, not to withdraw from acting properly and impartially.

As my interest in infectious disease control grew, and it became my professional specialism, I had dealings on a daily basis with fellow health professionals, patients, their contacts, and those otherwise implicated in the transmission of infection, especially commercial enterprises. The need for discretion, honesty, and the unalienable right of the patient or whoever to sanction what personal information should be shared by whom and to whom became instinctive. I should add that I was then involved in voluntary work with an organisation that had these principles at its core, so the limits were readily understood. Consequently, when drafting policy and procedure documents, one would be especially keen to emphasise the moments when the interests of others needed to be respected.

The gentle art of persuasion is a significant part of the work of EHPs, but the need to exercise all the principles of ethical behaviour and practice came to prominence when working in infectious disease control at a large, predominantly urban, borough council. This was the time when AIDS/HIV was inflicting its worst on the country, especially an increasingly marginalised local gay community. It is 1988, and we were beginning to see the impact of this seriously debilitating disease on our residents, some of whom were encountering serious difficulties in many aspects of their lives.

For those that can remember it, the appearance of the government's "Don't Die of Ignorance" campaign, in part intended to inform and reassure, served to increase the widespread vilification of those dealing with the consequences of infection, and promoted fear among the population at large. It was my, and my immediate colleagues, view, that the council's environmental health department could and should do something to counter the prejudice disrupting the lives of so many, but we needed some pointers as to how we might go about it.

The inspiration and encouragement to go further than my natural instincts would have come when attending a break-out session at the CIEH annual conference, where the London Borough of Hackney's response to the "AIDS crisis" was outlined. Explaining the Council's political commitment to the principles of equal opportunities, the presenter described how this had been applied to the unfounded fears of HIV infection through casual contact based on spurious "health and safety" concerns. By extension, this should, he explained, not be allowed to justify prejudicial attitudes and behaviour denying people services, employment and personal dignity. The Council's responsibility was for the public's health, and it had no right to pick and choose who should receive protection, even when personal doubts and fears were brought into focus.

In hindsight, this hour-long presentation indicated that ethical practice demands bravery to challenge the instinct for self-protection and the tendency to choose to do no more than is expected of a local government officer. A lesson learned was that going beyond the call of duty necessarily carries with it risks to the initiator, although the good that can be achieved outweighs the consequences of doing nothing. I was fortunate that my chief officer was sympathetic to the idea of an initiative, but we both knew that the greatest hurdle would lie in persuading Council members to get behind it in principle; moreover, to commit time and resources for doing something that flew in the face of central government's rhetoric at the time.

After a week or two considering how we should go about addressing the Environmental Services Committee and rehearsing the possible arguments, we might need to put forward in order to counter the anticipated "push-back", we briefed the Chair at a pre-meeting. It was heartening to receive her support for promoting a far more inclusive and empathetic message on AIDS/HIV. Nevertheless, there was a sense of foreboding as the item was introduced at the meeting and we outlined what we had in mind, foreseeing that the members would direct us to follow the Government's line. Their argument might be that by advocating a positive and caring attitude, the Council could find itself criticised as condoning what some saw at the time as unacceptable behaviour, such as homosexual lifestyles. There was also the issue of how this was affecting people needing blood transfusions, and anyone having unprotected sexual intercourse, which might include the Council's own staff.

Instead of objection, the Committee, and subsequently the Full Council, agreed to a series of informal talks to be made to all council staff on the nature of the infection, the means by which the virus could, and, could not, be transmitted, and the expectation that services should be provided to all regardless of their known or suspected HIV status. Thereafter, the same sort of presentation should be made available to the larger employers and community groups in the town. This was in response to calls received from managers of a number of hospitality outlets where gay male staff were receiving homophobic comments from their colleagues, in some cases their refusing to do shifts with the person.

Ahead of arranging the talks we had other matters to attend to that were more pressing. This included the difficulty of finding a location for a "day centre" for those living with AIDS/HIV, some of whom were visibly unwell and experiencing isolation through community prejudice. The "open meeting" proved lively, but it passed without too much anger being expressed and concluded positively once the two sides had expressed their hopes and fears. This meeting also served to bring us together with those who were active on the AIDS scene, one, the Chair of the local "Body Positive" group, invited me to attend their meetings on an "honorary" basis! This provided an insight into what many were going through and allowed me to properly reflect the difficulties they were experiencing. Above all, it allowed its members to see that the Council could be trusted.

Otherwise, a booklet was produced that sought to correct or qualify the remarks made in "Don't Die of Ignorance", and with the assistance of a technical officer with graphic skills, we designed posters for Christmas, New Year, and Valentine's Day that portrayed a list of things, including condoms, that might be put on a list of presents for each occasion. In hindsight these were clumsy, but the director of the Health Education Unit of the district health authority approved of them and saw several thousand laminated copies printed with adhesive backing. These were distributed around town in late Autumn for putting up in both male and female toilets.

The presentations proper got underway shortly afterwards. From the start, it was agreed that we would not shy away from any issue, even if such candour might offend. From memory, we never encountered anything other than a positive response, though there were moments when emotions ran high because of the pain being caused to predominantly young people. It concluded with an excellent video that featured Professor Michael Adler, then of University College London, that provided a neat conclusion to the scientific and factual aspects of transmission, diagnosis and treatment. The second half was a Q and A session which could be challenging as it was unstructured and freewheeling. In the early stages, the Chair of the Body Positive group would join this session, and he would be invited to pitch in when the inevitable question was raised about living with the disease. His involvement was not only of assistance in responding to questions in later talks but reassured us that we were on the right track when it came to the human side of those living with AIDS and HIV.

We conducted many, many sessions across a period of around 18 months, and though the ones undertaken in the Town Hall were logistically straightforward, I recall lugging a large TV/video player up flights of stairs and attending very early morning sessions for shift workers. Perhaps it's a shame now, as I write this, that I never made a note at the time as to the ethical implications of what we were doing. It seemed only natural to address the public health issue of the moment, drawing upon the ground-breaking work of the EHP at Hackney Council.

Upon reflection, this initiative did have major ethical implications to do with patient confidentiality when liaising with the local genitourinary medicine (GUM) clinicians, the need to challenge deep-rooted, personal feelings, and the right to express these at the expense of others. If we achieved anything it was to encourage the local authority to think "outside of the box" and recognise that a disease that had serious personal health implications should be considered a public health issue on which environmental health should not only give their support, but to lead.

As was the case in the early stages of the COVID-19 pandemic, many EHPs saw opportunities to initiate action to protect lives and livelihoods, and without direction from central government, assume the lead public health role, challenging the notion that personal health fell outside the purview of environmental health. Certainly, one of the most satisfying aspects of my career since retirement has been in canvassing local authority EHPs in connection with the COVID-19 pandemic. It was remarkable how they were prepared to break free from their "existential chains" and exercise their public health skills fearlessly for the public good, though mindful of the need to do so in an ethically prudent manner.

As something of a post-script, many years later I found myself working at the CIEH alongside the EHP from London who had inspired me back in 1988. At the time I was engaged in a review of the EHP qualification process, and having seen the revised curriculum and means of directing and assessing professional training introduced, sought to obtain recognition of the revised qualification package for the competencies it provided in public health. We chose to do this through the United Kingdom Public Health Register (UKPHR), who duly assessed the competences of an EHP coming through the new route against their criteria for admission to the register at Practitioner Level. It is interesting, and a source of personal embarrassment, that the only feature of the qualification process that was found wanting was the lack of explicit instruction in the ethical principles to be followed by practitioners, and how the professional body ensured that those qualifying in environmental health took these measures forward into practice. Such issues pertaining to ethical practice now seem to have been addressed, but it does suggest why this book will be a very welcome addition to the Routledge Environmental Health series.

Practitioner 6

In this case, the freelance EHP with expertise and experience in assessing housing conditions was approached by a firm of solicitors to provide expert evidence in an appeal against a Housing Act 2004 Notice and to assess the risk from a hazard identified by a local authority. The dilemma was that the owner of the property, the solicitor's client was based offshore in a tax haven, while legal this was something the EHP thought was morally wrong. On consideration of the papers and on the basis that the freelance EHP considered that risk had been overstated and there was evidence that the landlord had tried to address the risks. Given also that the owners had carried out some of the measures suggested by the local authority and that the terms of the notice were considered to be excessively onerous, the instruction was accepted. After a meeting of the local authority EHP and the freelancer after which there remained areas of professional disagreement, the case went to the First Tier Tribunal which varied the terms of the notice more in line with the views of the freelance EHP.

Unlike barristers there is no "cab rank rule" as set by the Bar Standards Board[8] that says a freelance EHP must take on any instructions where they are qualified to provide an expert opinion. However, again the published guidance from the professional body offered no real pointers on how to make that decision.

Comments and conclusions

Something that impinges on ethics and even professional ethics is that many EHPs in local government see themselves firstly as council officers rather than independent professionals responsible for public health. If there is no conflict (even if internalised by the person), this must question whether they are truly professional and members of a profession. If there is conflict, then the professional body must recognise this and play a role in securing resolution. Either way this has implications for protecting and even improving public health. As we have seen there is no guarantee that either law or public policy itself will always do that.

The scenarios given in this chapter highlight this. For some EHPs the route out of this is to leave local government. This is perhaps something that the professional body needs to examine closely. It is also apparent that for many EHOs/EHPs faced with an ethical dilemma they do not look to their professional body for support or advice. Guidance should more explicitly recognise that there are many EHOs/EHPs that have taken their skills and knowledge out of local government into private practise and other areas of the economy.

There is also a potential for conflict where a local authority EHO is asked to give evidence on behalf of the local authority when there is legal action concerning housing conditions in a local authority owned property. This could happen, for instance, under section 82 of the Environmental Protection Act 1990 or the Landlord and Tenant Act 1985 (as amended) or in the future where there is the involvement of the social housing regulator. This again is something where the professional body should provide support and guidance to ensure in such circumstances the EHO/EHP acts correctly.

Notes and References

1 Local Government Ombudsman (1993) Report on Investigation No.91/A/3602 Against London Borough of Newham.

2 In 1984 Baroness Vickers then a CIEH Vice President pointed out that tenants of local authorities were not protected by the legislation that applied to private tenants. This led to advice in DoE Circular 21/84 (W.O. No 42/84) relating to the Housing and Building Control Act 1984at para 117 that said that local authorities should make arrangements so that if the council tenant seeks the help of the EHO about conditions which would otherwise necessitate action by the Officer, the Officer "shall notify the Chief Housing Officer who should ensure the necessary remedial works are carried out within a reasonable time". Some years later this Circular was withdrawn as the result of further changes to legislation.

3 See also Ormandy D, and Burridge R, (1988) *Environmental Health Standards in Housing*, Sweet and Maxwell, London, p. 312.

4 See *Field v Leeds City Council* (2000) 32 HLR 618 CA, here it was held that although it was no bar that the expert the council wished to use was employed by them, it was important that they should be able to show that that the prospective witness 'has full knowledge of the requirements for an expert to give evidence before the court and that he is fully familiar with the need for objectivity'. Lord Woolf again stressed the desirability of such cases being conducted with only a single jointly instructed expert. *Bird v Hackney LBC* (2000) 15 June Central London County Court, Legal Action, July 2001, followed the *Field* case and the court rejected the expert evidence of the council employed surveyor as having insufficient objectivity.

5 Lipsky M, (1980) *Street Level Bureaucracy: Dilemmas of the Individual in Public Services*, Russell Sage Foundation. http://www.jstor.org/stable/10.7758/9781610447713

6 Hutter BM, (1986) An Inspector Calls: The Importance of Proactive Enforcement in the Regulatory Context. *The British Journal of Criminology*, Vol. 26, No. 2 (April), pp. 114–128. https://doi.org/10.1093/oxfordjournals.bjc.a047589

7 Hutter BM, (1989) Variations in Regulatory Enforcement Styles. *Law &Policy*, Vol. 11, No. 2, pp. 153–174.

8 The rule says that where the services of a named "authorised individual.........., and the instructions are appropriate taking into account the experience, seniority and/or field of practice of the named authorised individual" then, "the instructions must be accepted", irrespective of the identity of the client.

5 Final thoughts and concluding comments

To paraphrase George Bernard Shaw somewhat:

> The reasonable person (or professional) adapts themself to the world; the unreasonable one persists in trying to adapt the world to themself. Therefore, all progress depends on the unreasonable person.[1]

Being unreasonable does not mean being unethical, rather it might be seen that having a strong personal and professional ethic might lead one to being seen as unreasonable. As we saw in Chapter 1, Chadwick was not universally liked. It can also be seen as a call not to fear failure or disappointment for doing or saying what is right. Codes of Ethics for members of a profession perhaps need to reflect this.

Environmental health as a profession began with ethical purposes. It should now be the role of the professional body to ensure that members can act in their professional capacity without fear or favour. Developing and policing an effective code of ethics and conduct is considered vital for the future of the profession and for the health of future generations.

Professionals, particularly environmental health officers (or practitioners) (EHOs/EHPs), in their working environment encounter problems every day that should require them to check in with their moral code. Deliberately a polemic, this book has made the case for strengthening the role and recognition of EHOs/EHPs by having a code based on clear ethical principles and to which anyone professing to be an EHO/EHP (whether a member of a professional body or not) should be required to "sign up". An ethical code of practice binding on the membership is a prerequisite to being regarded as a profession. It should afford protection both to the practitioners and to their clients. It should also be available as a defence against pressures to distort professional judgements. This would be further strengthened if the profession was "regulated" and in local government the status of "proper officer" was defined accordingly.

It is apparent that, as Hardwick[2] has suggested, there are occasions where the desire to act ethically causes tensions with senior managers, whose priorities

DOI: 10.1201/9781003625339-5

may run counter to high ethical standards. For all that, findings from Hardwick's study indicated that environmental health was an ethically strong profession at that time. It was suggested that although ethical dilemmas do occur often, EHPs had developed their own ways of resolving these, though problems of conscience remained. Sometimes the ethical position was incompatible with what he saw as the rigid process of law, but rarely was the professional ethical code incompatible with personal moral principles and it was the latter which led to compliance with the code of ethics with which they were unfamiliar. That is perhaps not a reliable approach when tensions and pressures are increasing but also shows how there is a need for greater publicity for environmental health ethics.

It is clear that EHPs whether members of professional bodies or not, do have to make ethical decisions. At the highest level this could arise when, for example, a law is passed that they are required to enforce but which will harm public health – what if, for instance, smoking bans were reversed, or steps to improve air quality were reversed. It could be that a policy to improve human health harms or depletes the natural environment with habitat or species loss. Korthals[3] gave another example of numerous fast-food outlets on streets and the EHO might think the advertising of and sponsorship by these chains contributes to increasing obesity and other diseases such as type 2 diabetes, cardiovascular disease, and cancer of the intestines but "one of the most powerful food companies, thinks otherwise and acts accordingly". In these circumstances, it is not sufficient for the EHP to focus only on food hygiene, their traditional area of work, in such premises. As Korthals says, "environmental risks to health are for the most part not equally distributed, so that rich people tend to live in healthier environments than do poor people". Perhaps it is the collective role of the professional bodies to pursue the wider health issues and promote the holistic approach to health.

Environmental health is part of public health

It has been suggested previously that medical ethics are of no assistance when developing public health ethics and this book has worked from the basis that EHPs are part of the public health workforce. That said, outside Scotland this does not always seem to be explicitly accepted and so it needs to be highlighted more often.[4] Public health involves a population emphasis and prevention of ill health by identifying threats to health (and safety), this contrasts with the essentially individual focus of medical diagnosis treatment and relief of suffering. The medical approach rarely looks at the environmental factors influencing ill health or injury. It rarely considers societal factors and the social determinants of health. Environmental health ethics, therefore, should not be seen as separate from public health ethics. Indeed, the professional bodies should be providing greater insight into ethics not just professional ethics as a code of behaviour important as that is, and this should be inculcated into environmental health students.

As part of the public health workforce, and with public health agendas aimed at addressing health inequalities and well-being, there is an ethical obligation that also implies a political morality or integrity. Ethics is an integral part of public health decision-making and should be incorporated into public and environmental health policy development and decision-making.

Research

All undergraduate students need ethical approval for any research involving gathering or holding data from human participants in any form. Most readers will be familiar with the requirements of ethics committees in universities. These are responsible for reviewing ethics applications to ensure that adequate consideration has been given to the ethical aspects of the research project. The aim is to reduce the potential for harm and upset to the participants. The research should aim to maximise benefit for individuals and society and minimise risk and harm and respect the rights and dignity of individuals and groups. In common with what has been discussed earlier, research should be conducted with integrity and transparency, and if conflicts of interest cannot be avoided, they should be made explicit, and research should remain independent. In research, lines of responsibility and accountability should be clearly defined. This might not be so obviously relevant in the local government sphere, but in a commercial or consultancy setting, remains relevant. In research, there is always the need to consider risk versus benefit and for the EHO/EHP that is particularly true, for example, when considering the work necessary to address a hazard or hazards in a dwelling or in the workplace.

Research is in any event a part of a professional's career and is a more constructive approach to professional development than some of the ways of accruing continuing professional development (CPD) hours. This is something that is within the remit of professional bodies to support, but while they may undertake some of their own research they often do little to support members in this area. That is a sad reflection where this is the case, as this is a route for highlighting environmental health work that is underused.[5] Open sharing of research outcomes is an important part of professional training and development. This encourages collegiate activity, common understanding, and better decisions and policy development.

Professional codes in today's world

As has been made clear in this book, a robust code of ethics and its application is a prerequisite of any profession. It is not always clear to what extent ethics are promoted and enforced within the different professions. It is not sufficient for only lip service to be paid to an ethical approach. Having a strong ethical code does not prevent members behaving unethically so the code must be supported by a robust system of enforcement that is also transparent. It is also worth making mention again of the NEHA approach, requiring a member to print and sign a document indicating they acknowledge, accept, and agree to abide by the code of ethics. Similarly in Canada, the code requires that all persons applying for or renewing membership in CIPHI to acknowledge the CIPHI code of ethics and agree to conduct themselves in accordance with the fundamental principles and ethics in the code. This approach brings it home to members of their ethical responsibilities. It is also worth mentioning that the documents for CIPH include a form for complaints under the code.

For the EHO or EHP, it is essential that they adhere to a professional code and understand ethics. This is because they deal with members of the public daily whether at home, at work, or at leisure where they make decisions that affect the lives of individuals and communities. It is therefore fundamental that there should be an ethical basis for any advice given and any responses to questions need to be trusted as honest in purpose founded on evidence and professional knowledge, and without prejudice or unconscious bias. Professional independence is also essential just as it is in the world of research. This should be reflected in advice given within the employing organisation and in policy formulation or development.

It can be argued that being an EHP is inherently an ethical profession, but while the potential is there, it is only true in part. A member of the profession needs to understand ethics and their ethical responsibilities. There is a need for regular reminders. Knowledge and understanding of ethics should not be taken as read. Without an understanding of ethics, an EHO/EHP risks becoming a bureaucrat interested in process rather than having the objective of improving or safeguarding public health. That is, it is too easy to lose sight of their purpose. Any action taken should not just be assumed to improve the health or safety of the community served but should be demonstrable wherever possible.

At the outset of writing this book it was thought that strong professional ethics would also promote the profession and increase entry, but perhaps this was perhaps too optimistic. However, it would not hinder recruitment and can be used to demonstrate the value of the profession, something that is important given the nature of the work of EHOs/EHPs. It should also be a "selling point" that there are strong professional ethics. As part of that, members of the profession and its professional body itself should be seen to be working to improve public health and address health inequities.[6] This then implies more than just paying lip service to professional ethics and codes of behaviour. It means going beyond the technocratic or mechanical application of technical knowledge. At the Chartered Institute of Environmental Health (CIEH), the Professional Standards Framework could make a great deal more of ethics than it does,[7,8] but that is certainly true of most if not all the environmental health professional bodies considered when writing this book. It should also be recognised that professional ethics is "not a one-way street" and professional bodies should be giving support to members on ethical issues.

Recent changes in local government as the result of the latest local council elections should also be considered. Some councils are now controlled by a party that mimics the approach seen in the USA.[9] These changes further exemplify what has been said earlier about policies that can conflict with ethics. Reports suggest a threat to professionals in local government.[10] As a matter of ethics, any policy to address health threats to populations or sections of society should include a health equity assessment. It is EHOs/EHPs in a local authority or commercial company who should be equipped to undertake this. However, what we might consider to be an ethical approach to an environmental health problem may not be so to these councillors now in control of some councils. Looking at what has happened in the USA and attacks on what is called diversity, equity, and inclusion (DEI), then it is likely they will not support action to address health inequity. While it might be early days, this is something that environmental health professional bodies in the UK need to consider as to how they deal with this new issue.

It has been alleged that EHOs/EHPs have used "the ways and means Act" or "section bluff" of "informal action" to resolve perceived problems even where there is legislation to support action. This may be an attempt to resolve a problem that can arise because of legitimate concerns and worries, by using diplomacy. One might even argue this is done for ethical reasons or motives. That, however, also leads to problems of inconsistency and unfairness. It is also possible that the linking of health and well-being has given rise to problems with EHOs/EHPs seeing many problems including social problems as "health" or "public health" problems. The environmental health profession and its professional bodies need to be aware of this possibility. This can be problematic when it can be managerially or politically convenient to give additional responsibilities to environmental health teams. That does not mean EHOs/EHPs should not be working with other agencies and professions to resolve such problems.

In a similar vein, there is the question of the course of action for the professional when local practice is not to use powers in law intended to protect the health and safety of occupiers. This may be because of failures of other parts of the organisation or a lack of confidence in other parts of the local authority. For example, in the case where the guidance on prosecuting would be met and it would be in the public interest, the lack of confidence in the administrative, legal, or financial support means that neither the route of prosecution nor civil penalty is taken.

It is not known if any EHPs/EHOs have ever contacted their professional body where they have felt the need to put the interests of the communities they serve and public health first before those of any colleague or organisation, including their employer. While for the CIEH their members "shall take swift and appropriate action if they become aware that ...... the actions of an organisation, may compromise the health of the public". This is an example of where there is guidance but also gaps. It is not clear what action the member should take in these circumstances and how the CIEH will handle this. If a member is seeking to behave ethically, but this causes conflict with their employer or the person to whom they are contracted, it is incumbent on their professional body to set out how this should be dealt with. If it is expected that a member of a professional body should seek advice from an appropriate colleague when they are unsure how to deal with a particular situation, then the criteria for "appropriateness" need to be set out including their independence. Again, the professional body should be in a position to provide advice and investigate further if necessary. In that regard, codes of ethics should also be seen as part of services to and support for members.

Where there are any perceived ethical conflicts, these should be reported to the professional body. Knowledge and understanding of such conflicts will help the professional body improve their code and advice. Some professional bodies contain a whistleblowing policy within their codes; this is something that environmental health professional bodies should also include in their approach to ethics and include advice for its members on this.

There is a general requirement in professional codes that members should decline to act if they lack the knowledge, skills, experience, or authority to undertake a piece of work and should not give advice where they have reason to believe they lack the professional knowledge and skill at the level required. It would clearly be unethical to act beyond one's competence, but it also requires high-quality CPD and keeping up to date with developments. This means that the professional body should be rigorous in the quality of acceptable CPD hours and its enforcement of CPD provisions.

It should be a matter of course that a professional code be well publicised including within the profession itself. That said, most codes focus on expected behaviour and less on wider ethics and tend not to include what the professional body can do to support members on matters of ethical interest or where they encounter ethical dilemmas. From what we have found and written in the earlier chapters, any consideration of professional ethics and associated codes should take better account of the need for support for members on questions of ethics. Professional codes often do not explicitly set out what a member should do, or where the members should turn when they believe policies and practices are incompatible with ethics or compromise their personal morals. In the case of EHPs, this could be where the approach is only to individualise problems rather than look at the population or community level. This might be reflected in different aspects of environmental health work. For example, a more strategic approach might be taken when it comes to food hygiene inspections, but there is too often a reliance on individual complaints when it comes to housing conditions. At the same time judgements are being made on how the law should or should not be used. A greater understanding of ethics would help the decision-making when it comes to use of the appropriate legal basis for action to resolve a problem. We asked ourselves whether professionally the codes of Ethics help answer such questions and tensions between courses of action. In most cases they do not. In Chapter 4, we saw how it was not possible to turn to managers for assistance when ethical issues arose, so this is a role for the professional body.

The professional bodies considered go into varying levels of detail on the behaviours expected as part of their code of ethics and to meet the obligations of their code. Even with a brief outline of the ethical principles which again vary across organisations, it is a moot point whether increased detail improves or contributes more to ethical behaviours on the part of that profession. There is a question as to whether such a detailed set of behaviours achieves the aim. Without adequate publicity and reinforcement any code will have reduced impact on behaviour. Simpler and shorter codes could be as effective if supported by reinforcement and guidance. There is an argument for re-assessing such codes and notions of fitness to practise to ensure there is no "mission creep" but also that they are capable of effective implementation. The approach of the CIPD in working with the Professional Associations Research Network (PARN) and academic institutions' units specialising in professional ethics is a sensible approach that should be copied when updating codes of ethics and fitness to practise.

Fully qualified EHOs/EHPs are equipped with the skills and knowledge to safeguard and improve the health and lives of the public whether employed in the public sector or private enterprises. These are transferrable skills that can be better utilised when there is a focus on just one aspect of environmental health work such as food safety or unhealthy housing even if as a "registered" practitioner. On top of this it would make sense for all the United Kingdom to have the same legislation as in Scotland which requires the appointment of an EHO who meets the requirements of the professional body. In other countries, the role and functions of EHOs/EHPs are set out in the law. Regrettably this is unlikely to happen in the UK outside Scotland. At the same time and in any event the professional bodies have a role to play in providing leadership and demonstrating the value of environmental health professionals to society.

A licence to practise usually implies a controlled entry with professional ethics underpinning this. Looking at the situation in several countries and even at Building Control since Grenfell, it is worth considering whether it would be better if the code of ethics for environmental health had some statutory backing (although for building control that is a result of the tragedy). Or at least that there should be some greater control similar to Scotland.

In the absence of any statutory requirement to employ EHOs as "proper officers", strong and highly publicised enforced codes of professional ethics should encourage potential employers to only recruit members of the CIEH. This does not directly increase the number of new graduates nor of them progressing to becoming fully qualified EHOs/EHPs given they must compile a portfolio[11] of practice that focuses very much on the technical aspects of environmental health and undergo a "professional discussion". Financial pressures may not encourage local authorities to increase support for training. While not a matter of ethics, this is not good for the future. Some means must be found for addressing this problem, and the new Growth and Skills Levy could be a potential route, as could some top slicing of income from licensing and civil penalties to support the training of EHOs/EHPs to deal with unhealthy housing. Unfortunately, while the Government has increased funding for apprenticeships, environmental health does not appear to be seen as a priority sector and as the CIEH President has said "environmental health professionals make a huge difference to the public's health and wellbeing, and their work is only getting more important".[12] Many of the current generation of EHOs/EHPs will have received free training and been supported by employers towards gaining their environmental health qualification. The profession should be looking at more and imaginative ways that reduce the cost of entry borne by individuals and lobbying to garner political support. A robust code of ethics can help in that regard.

The WHO has said that the numbers of EHOs should be sufficient to exercise adequate surveillance over health-related environmental conditions and monitoring activities. A CIEH survey covering England[13] and referred to in the latest edition of *Clay's* indicated that in 2020 there was a ratio of six fully qualified EHO/EHP employed for every 100,000 of the population[14] (assuming the higher figure for the estimated full-time equivalents). Whether that is adequate given the range of environmental health problems is a moot point and there is no indication that the ratio has increased since then. It has been argued that just to address problems in the private rented housing sector a ratio of 5 EHOs per 10,000 PRS households is the minimum required to effectively regulate a PRS market.[15] This highlights the inadequate numbers of EHOs currently working in local authorities. Employing properly qualified environmental health professionals operating within a robust ethical code should also constitute a defence in the event of an inquiry or litigation.

It is questionable whether in the UK, registration confers sufficient advantage when it comes to employment. Under the new CIEH regime maintenance of registration also requires maintenance of CIEH membership and so adherence to the code of ethics. Yet many who registered under the previous regime may choose not to pay the fee to be included on the new CIEH register. If not members of the CIEH, the code of ethics cannot apply to them. This is most certainly a gap that is another reason why reform of the Environmental Health Registration Board (EHRB) might have been a better approach than a completely new and separate approach to registration. It might be that this can only now be addressed by legislation unless it is possible to reconstitute a reformed EHRB. For chartered status, this should incorporate a clear ethical requirement which is beyond the knowledge, skills, and behaviour required specifically for the profession.

It is an ethical question as to the impact of financial considerations, and whether at times of economic strictures the desire to maximise membership fees compromises regulation of a profession and the ethics of the regulator of the profession. This might also point to the need for an independent registration system. Looking around the world there are different approaches to registration and legal status. There is no evidence to suggest that regulation of a profession is compromised in this situation. However, if there is little publicity or reinforcement of the codes, they will not be well used with few complaints or referrals and disciplinary action will be rare even if reported publicly. Nevertheless, a conflict of interest can be perceived, and on that basis any system of registration would be better to have some independence or at least subject to independent oversight and it would be better if registration and fitness to practise (including ethics) was not dependent on membership of the professional body. Such registration should not be for life and the professional body should have the ability to make representations to have a registrant struck off or suspended for serious breaches of the code of ethics and fitness to practise or other malpractice by any EHP/EHO.

Conclusion

Ethics can be at a philosophical level and at the professional level. This book has largely considered the latter but has not ignored the former.

EHPs (or EHOs) in their every-day work may not think they are making ethical decisions or have need to be aware of ethical issues, but they are, whether it is fashionable to think in those terms or not. As an example, the fairness or justice approach to ethics has its roots in the teachings of Aristotle in that favouritism and discrimination are unjust and wrong and fairness requires consistency in the way people are treated. This also means avoiding or recognising and addressing unconscious bias. Thus, decisions on whether to take a "formal" (serving a notice) or "informal" approach to securing compliance or changes in behaviour is a matter of ethics.

The ethics of a profession are normally dictated by the professional body, but this is not consistent around the world. This often depends on the legal status of the profession and how it is regulated. In the UK in the main, regulation of a profession delegated to the professional body and that is true of the environmental health profession. Whatever regulatory role any environmental health association or institute might have, they should also be able to provide support for members on ethical matters.

It must be incumbent on the professional body that as well as setting standards of behaviours for its members it should itself act ethically. This might entail having in place an ethics committee, or something similar within its governance structure. They should also consider developing ethical frameworks (beyond sets of behaviours) and models to assist members with ethical decision-making. This would help them explain the ethical basis for actions.

This also requires that consideration of ethics should be part not only of training but clearly part of the approach to in-service training and CPD.

This leads to one final comment. Fully qualified environmental (or public) health officers are trained to have a wide range of knowledge and skills that are transferrable whether they later choose to specialise in one area of work or not. Registered practitioners specialising in one aspect from the outset are not and there is no reference to ethical issues in the published routes to qualification. Though expedient this should be seen as a short-term approach not just for the profession but for society (as exemplified by issues raised during the COVID pandemic). It should be the aim for such practitioners registered in one aspect of environmental health to become fully qualified as EHOs/EHPs and membership requirements should support and reflect this. To encourage membership of the professional body and be subject to their codes, it would also make sense for student EHPs to have free membership of their professional body given the other pressures they face en route to qualification.

The profession should present an opportunity to carry out worthwhile work that profits society. This should remain an ambition even in these cynical times. It is hoped that this book will stimulate debate within the environmental health profession wherever located and whatever the nature of work whether in private practice or within government at whatever level.

Notes and References

1 This is based on Geoge Bernard Shaw's (1903) Maxims for Revolutionists which was an appendix to his play Man and Superman: A Comedy and a Philosophy.

2 Hardwick, M, (2015) Understanding the Ethical Positioning of Environmental Health Officers in Work-Based Situations: A Study Illustrating the Use of a Phenomenological Approach. *Journal of Environmental Health Research*, Vol. 15, No. 1 (December), pp. 87–97.

3 Korthals, M, (2011) Ethics of Environmental Health, in Chadwick R, ten Have H, and Meslin EM, Eds, *The SAGE Handbook of Health Care Ethics: Core and Emerging Issues*, pp. 413–426. Sage Publications. https://doi.org/10.4135/9781446200971.N34

4 It is worth noting that CIEH members are entitled to free reciprocal membership with the Faculty of Public Health.

5 This was the subject of a CIEH Task Group report in 2012Task Group and among other things commented that "'research' should not reside solely in the positivist scientific domain but may incorporate areas such as phenomenological experience retrospective analysis of events and other approaches not necessarily randomised control trials" and thus has a wider meaning than might commonly be assumed.

6 Health inequities are systematic differences in the health status of different population groups. These inequities have significant social and economic costs both to individuals and societies see https://www.who.int/news-room/facts-in-pictures/detail/health-inequities-and-their-causes

7 CIEH (2018) *CIEH Professional Standards Framework – Professionalism in Practice; Guide*. Available to members at https://www.cieh.org/media/2225/cieh-professional-standards-framework-guide.pdf

8 At the time of writing the CIEH is about to launch a review of the professional standards framework https://www.cieh.org/professional-development/our-professional-standards/

9 https://www.localgov.co.uk/Reform-UK-launches-DOGE-teams-to-investigate-council-finances/62467

10 https://www.thelondoneconomic.com/politics/who-are-new-reform-mayors-sack-diversity-officers-392762/ and https://www.prospectmagazine.co.uk/politics/democracy/government/local-government/69932/what-to-expect-from-reform-run-councils-nigel-farage

11 This includes the need to be supported by a workplace supervisor assigned by the employer with a minimum of six months spent working on the portfolio that covers interventions in all areas of environmental health work.

12 CIEH Press Release of 2nd June 2025 CIEH welcomes Government announcement on apprenticeships but wants greater focus on environmental health at https://www.cieh.org/news/press-releases/2025/cieh-welcomes-government-announcement-on-apprenticeships-but-wants-greater-focus-on-environmental-health/

13 CIEH (2021) *Environmental Health Workforce Survey Report: local authorities in England*, CIEH, London.

14 Taking the population of England as 56.5 million and the workforce survey estimated between 3,240 and 3,360 full time equivalent fully qualified EHOs working for local authorities.

15 https://metastreet.co.uk/blog/where-have-all-the-housing-ehos-gone.html

Index